Collins

AF776266

Greater Depth Maths

Build curiosity and confidence

TEACHER GUIDE LOWER KEY STAGE 2

Herts for Learning

William Collins' dream of knowledge for all began with the publication of his first book in 1819. A self-educated mill worker, he not only enriched millions of lives, but also founded a flourishing publishing house. Today, staying true to this spirit, Collins books are packed with inspiration, innovation and practical expertise. They place you at the centre of a world of possibility and give you exactly what you need to explore it.

Collins. Freedom to teach.

Published by Collins
An imprint of HarperCollins*Publishers*
The News Building
1 London Bridge Street
London
SE1 9GF

HarperCollins*Publishers*
1st Floor, Watermarque Building
Ringsend Road
Dublin 4
Ireland

Browse the complete Collins catalogue at

www.collins.co.uk

10 9 8 7 6 5 4 3 2 1

ISBN 978-0-00-845491-3

British Library Cataloguing-in-Publication Data
A catalogue record for this publication is available from the British Library.

Authors: Charlie Harber, Rachael Brown, Nicola Adams, Laura Dell
Publisher: Lizzie Catford
Product manager: Letitia Luff
Editorial management: Oriel Square
Cover designer: Amparo Barrera
Cover photograph: Bubutu/Shutterstock
Typesetter: Jouve India Pvt. Ltd.
Internal illustrations: Jouve India Pvt. Ltd.
Production controller: Alhady Ali
Printed and bound in the UK using 100% Renewable Electricity at CPI Group (UK) Ltd

FSC
www.fsc.org
MIX
Paper from responsible sources
FSC™ C007454

This book is produced from independently certified FSC™ paper to ensure responsible forest management.

For more information visit: www.harpercollins.co.uk/green

The publishers gratefully acknowledge the permission granted to reproduce the copyright material in this book. Every effort has been made to trace copyright holders and to obtain their permission for the use of copyright material. The publishers will gladly receive any information enabling them to rectify any error or omission at the first opportunity.

Digital downloads available at:

www.collins.co.uk/HertsforLearning/downloads

Contents

Getting Started Guide **4**

What is greater depth, and what does it look like? 4

Collins Greater Depth Maths 4

Lower Key Stage 2 tasks and the National Curriculum 4

The components 9

The key features of the resource 10

How to use the Teacher Guide 12

How to use the Pupil Resources 18

Assessment rubric 20

Task 1: Who Am I? **24**

Task 2: Place Value Pebbles **34**

Task 3: Simple Subtraction **44**

Task 4: X Marks the Spot **54**

Task 5: Triple Trick **63**

Task 6: Marching Elephants **72**

Task 7: Fair and Square **81**

Task 8: Beat the Calculator **89**

Task 9: Stablising Snowflakes **98**

Task 10: Queen Bee **107**

Task 11: Balancing Act **117**

Task 12: Perfect Poggle Pens **126**

Task 13: Shapes Within Shapes **135**

Task 14: L Plates **144**

Task 15: Blown Away **152**

Getting Started Guide

What is greater depth, and what does it look like?

Greater-depth learning is more than cognitive content; it is complex and pays attention to mathematical structure and what lies beneath a concept or procedure. A child working at greater depth is ready and able to explore mathematical concepts in more detail. Key characteristics include being curious and asking interesting mathematical questions, making connections and identifying patterns, collaborating with others effectively, communicating their ideas with increasing clarity, being creative with their responses, and considering arguments and providing counter-arguments with proof.

Collins Greater Depth Maths

The **Collins Greater Depth Maths** resources have been written by an experienced team of maths practitioners. Features of these resources include:

- high-quality mathematical investigations that spark curiosity;
- professional development embedded to increase teacher confidence;
- ready-to-teach sequences;
- 'stumble support' to provide inclusive greater-depth opportunities for all children.

The resources are provided for three phases of the primary curriculum: Key Stage 1, Lower Key Stage 2 and Upper Key Stage 2. There are 15 tasks for each phase, all fully explained and supported.

Each task within a phase has a comprehensive set of resources. The **Teacher Guide** provides detailed guidance showing teachers how to teach and facilitate each task, with different options to suit classroom needs. Each task comprises a set of component elements and could take a full lesson or several lessons. Children are given extensive opportunities to explore the mathematical concept under investigation and to discuss and share ideas.

The innovative **Pupil Resources** contain task-specific photocopiable pages that support children at each stage of a task, to develop independence.

Lower Key Stage 2 tasks and the National Curriculum

The charts on pages 5–8 list the Lower KS2 tasks and their relevant National Curriculum objectives.

All tasks can be done in either Year 3 or Year 4, but to give some guidance, the chart suggests which Year group could focus on a given task. The allocation between Years 3 and 4 ensures good coverage of different maths topics across the Year groups. To provide teachers with some flexibility, three of the tasks are allocated to both Years 3 and 4.

In the **Teacher Guide**, a small icon indicates the Year group for which each task is suggested.

Year 3 Year 4

	Year 3 National Curriculum	Year 4 National Curriculum
Number and place value		
1 **Who Am I?** Year 3	• Recognise the place value of each digit in a three-digit number (hundreds, tens, ones). • Compare and order numbers up to 1000. • Identify, represent and estimate numbers using different representations. • Read and write numbers up to 1000 in numerals and in words. • Solve number problems and practical problems involving these ideas.	• Recognise the place value of each digit in a four-digit number (thousands, hundreds, tens, and ones). • Order and compare numbers beyond 1000. • Identify, represent and estimate numbers using different representations. • Solve number and practical problems that involve all of the above and with increasingly large positive numbers.
	Links to White Rose: Year 3 Autumn – Number: Place Value	
2 **Place Value Pebbles** Year 4	• Identify, represent and estimate numbers using different representations. • Solve number problems and practical problems involving these ideas.	• Identify, represent and estimate numbers using different representations. • Solve number and practical problems that involve all of the above and with increasingly large positive numbers.
	Links to White Rose: Year 4 Autumn – Number: Place Value	
Addition and subtraction		

3 **Simple Subtraction**
Year 3
	• Add and subtract numbers mentally, including:
	  • a three-digit number and ones;
	  • a three-digit number and tens;
	  • a three-digit number and hundreds.
	• Add and subtract numbers with up to three digits, using formal written methods of columnar addition and subtraction.
		• Add and subtract numbers with up to four digits using the formal written methods of columnar addition and subtraction where appropriate.

	Links to White Rose: Year 3 Autumn – Number: Addition and Subtraction	
4 **X Marks the Spot** Year 4	• Add and subtract numbers with up to three digits, using formal written methods of columnar addition and subtraction. • Estimate the answer to a calculation and use inverse operations to check answers. • Solve problems, including missing number problems, using number facts, place value, and more complex addition and subtraction.	• Add and subtract numbers with up to four digits using the formal written methods of columnar addition and subtraction where appropriate. • Estimate and use inverse operations to check answers to a calculation. • *Pupils continue to practise both mental methods and columnar addition and subtraction with increasingly large numbers to aid fluency.* [non-statutory guidance]
	Links to White Rose: Year 4 Autumn – Number: Addition and Subtraction	

		Year 3 National Curriculum	Year 4 National Curriculum
Multiplication and division			
5	**Triple Trick** Year 3	• Recall and use multiplication and division facts for the 3, 4 and 8 multiplication tables. **Addition and subtraction** • Solve problems, including missing number problems, using number facts, place value, and more complex addition and subtraction.	• Recall multiplication and division facts for multiplication tables up to 12 × 12.
		Links to White Rose: Year 3 Autumn – Number: Multiplication and Division	
6	**Marching Elephants** Year 3 or 4	• Recall and use multiplication and division facts for the 3, 4 and 8 multiplication tables.	• Recall multiplication and division facts for multiplication tables up to 12 × 12. • Use place value, known and derived facts to multiply and divide mentally. • Solve problems involving multiplying.
		Links to White Rose: Year 3 Autumn – Number: Multiplication and Division Year 4 Autumn – Number: Multiplication and Division	
7	**Fair and Square** Year 4	• Recall and use multiplication and division facts for the 3, 4 and 8 multiplication table.	• Recall multiplication and division facts for multiplication tables up to 12 × 12. **Measurement:** • Find the area of rectilinear shapes by counting squares.
		Links to White Rose: Year 4 Autumn – Number: Multiplication and Division	
8	**Beat the Calculator** Year 4	• Recall and use multiplication and division facts for the 3, 4 and 8 multiplication tables. • Solve problems, including missing number problems, involving multiplication and division, including positive integer scaling problems and correspondence problems in which *n* objects are connected to *m* objects.	• Recall multiplication and division facts for multiplication tables up to 12 × 12. • Solve problems involving multiplying and adding, including using the distributive law to multiply two-digit numbers by one digit, integer scaling problems and harder correspondence problems such as *n* objects are connected to *m* objects.
		Links to White Rose: Year 4 Spring – Number: Multiplication and Division	

	Year 3 National Curriculum	Year 4 National Curriculum
Fractions		
9 **Stabilising Snowflakes** Year 3	• Recognise and show, using diagrams, equivalent fractions with small denominators. • Add and subtract fractions with the same denominator within one whole (for example, $\frac{5}{7} + \frac{1}{7} = \frac{6}{7}$).	• Add and subtract fractions with the same denominator. • Recognise and write decimal equivalents of any number of tenths or hundredths.
	Links to White Rose: Year 3 Spring – Number: Fractions	
10 **Queen Bee** Year 3 or 4	• Recognise, find and write fractions of a discrete set of objects: unit fractions and non-unit fractions with small denominators. • Recognise and show, using diagrams, equivalent fractions with small denominators. • *Continue to recognise fractions in the context of parts of a whole, numbers, measurements, a shape, and unit fractions as a division of a quantity.* [non-statutory guidance]	• Solve problems involving increasingly harder fractions to calculate quantities, and fractions to divide quantities, including non-unit fractions where the answer is a whole number. • Recognise and show, using diagrams, families of common equivalent fractions. • Recognise and write decimal equivalents to $\frac{1}{4}$, $\frac{1}{2}$, $\frac{3}{4}$.
	Links to White Rose: Year 3 Spring – Number: Fractions Year 4 Spring – Number: Fractions	
Measurement		
11 **Balancing Act** Year 3	• *The comparison of measures includes simple scaling by integers (for example, a given quantity or measure is twice as long or five times as high) and this connects to multiplication.* [non-statutory guidance] **Multiplication and division:** • Solve problems, including missing number problems, involving multiplication and division, including positive integer scaling problems and correspondence problems in which *n* objects are connected to *m* objects.	**Multiplication and division:** • Solve problems involving multiplying and adding, including using the distributive law to multiply two-digit numbers by one digit, integer scaling problems and harder correspondence problems such as *n* objects are connected to *m* objects.
	Links to White Rose: Year 3 Summer – Measurement: Mass and Capacity	
12 **Perfect Poggle Pens** Year 4	• Measure the perimeter of simple 2D shapes.	• Measure and calculate the perimeter of a rectilinear figure (including squares) in centimetres and metres. • Find the area of rectilinear shapes by counting squares. **Multiplication and division:** • Recognise and use factor pairs and commutativity in mental calculations.
	Links to White Rose: Year 4 Spring – Measurement: Area	

	Year 3 National Curriculum	Year 4 National Curriculum
Geometry – properties of shapes		
13 **Shapes Within Shapes** Year 3	• Draw 2D shapes and make 3D shapes using modelling materials; recognise 3D shapes in different orientations and describe them. • Recognise angles as a property of shape or a description of a turn. • Identify right angles, recognise that two right angles make a half-turn, three make three-quarters of a turn and four a complete turn; identify whether angles are greater than or less than a right angle. • Identify horizontal and vertical lines and pairs of perpendicular and parallel lines.	• Compare and classify geometric shapes, including quadrilaterals and triangles, based on their properties and sizes. • Identify acute and obtuse angles and compare and order angles up to two right angles by size. • Identify lines of symmetry in 2D shapes presented in different orientations. • Complete a simple symmetric figure with respect to a specific line of symmetry.
	Links to White Rose: Year 3 Summer – Geometry: Properties of Shape	
14 **L Plates** Year 4	• Draw 2D shapes. **Measurement:** • Measure, compare, add and subtract: lengths (m/cm/mm). • Measure the perimeter of simple 2D shapes.	• Compare and classify geometric shapes, including quadrilaterals and triangles, based on their properties and sizes. **Measurement:** • Measure and calculate the perimeter of a rectilinear figure (including squares) in centimetres and metres.
	Links to White Rose: Year 4 Autumn – Measurement: Length and Perimeter	
Statistics		
15 **Blown Away** Year 3 or 4	• Interpret and present data using bar charts, pictograms and tables. • Solve one-step and two-step questions [for example, 'How many more?' and 'How many fewer?'] using information presented in scaled bar charts and pictograms and tables.	• Interpret and present discrete and continuous data using appropriate graphical methods, including bar charts and time graphs. • Solve comparison, sum and difference problems using information presented in bar charts, pictograms, tables and other graphs.
	Links to White Rose: Year 3 Spring – Statistics Year 4 Summer – Statistics	

The components

There are three components for each task:

- **Teacher Guide**
- **Teacher's Presentation PowerPoint slides**
- **Pupil Resources.**

Teacher Guide

The **Teacher Guide** gives detailed teaching pathways for each task, including:

- guidance on key mathematical ideas;
- examples of age-appropriate children's responses;
- advice on how to model the identified strategies;
- prompts and sentence stems to scaffold children's learning and maintain engagement;
- word banks to deepen and focus vocabulary.

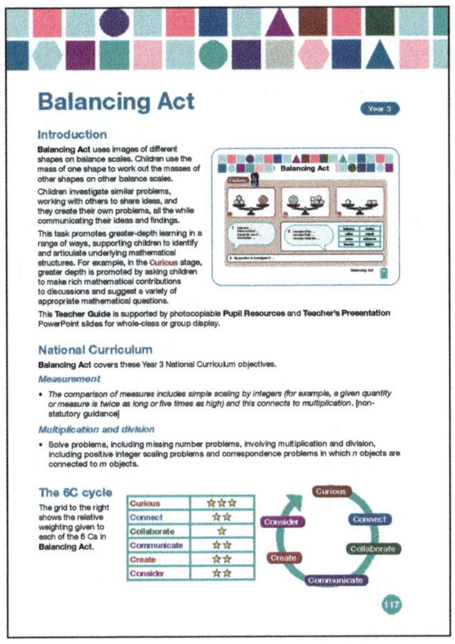

Teacher's Presentation PowerPoint slides

PowerPoint slides are provided, to enhance whole-class teaching and discussion. Some of the slides include notes providing 'stumble support' suggestions, that may help you to scaffold children in their problem-solving thinking and enable access for children who are less confident.

Where a printable handout might be useful, these are provided as black-and-white slides.

There are also slides of the **Pupil Resources**, should you wish to display these.

Pupil Resources

The **Pupil Resources** comprise printable pages for classroom distribution. The pages show visual representations of the problem being investigated, with prompts and word banks to help children discuss and develop their ideas. For each task there are four pages of **Pupil Resources**: one page each for the **Full** and the **Express Curious** tasks (see below for an explanation of **Full and Express Pathways**), and two pages for the remaining 5C tasks which are common to both pathways. It is suggested that teachers photocopy the pages and laminate them to create a reusable card-like resource, giving children freedom and flexibility to explore the tasks in the different Cs.

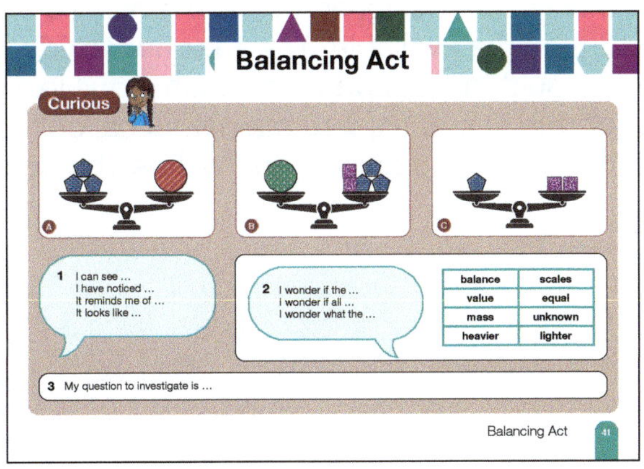

The key features of the resource

The 6C cycle

The 6Cs represent aspects of the desired characteristics of greater-depth learning: to be curious, to connect mathematical concepts, to collaborate with one another, to communicate their ideas and findings, to create their own examples and to consider and review their work and the work of others.

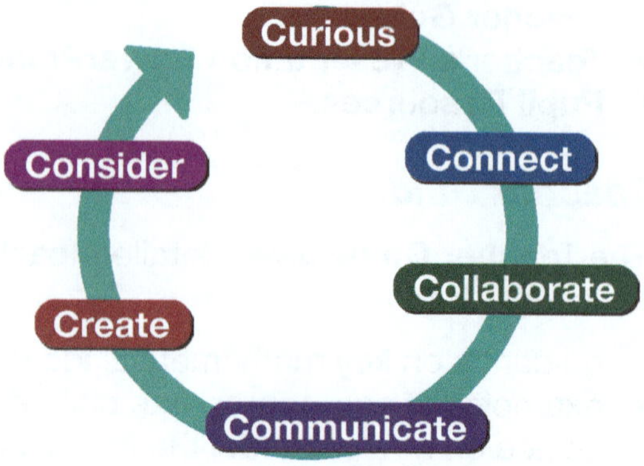

- **Curious:** promoting and encouraging curiosity in children.
- **Connect:** supporting children to connect their immediate observations to their past learning.
- **Collaborate:** developing meaningful collaboration between peers as a worthwhile skill, not just in mathematics but in everyday life.
- **Communicate:** supporting children to communicate mathematically with precision, clarity and respect.
- **Create:** fostering further curiosity and creativity.
- **Consider:** reflecting on mathematical learning experiences.

Problem-solving heuristics

Children are often good at applying a single strategy and repeating it, without necessarily understanding why it works, or how it might be applied in different situations. The tasks in **Greater Depth Maths** encourage the use of a range of problem-solving heuristics (strategies), and promote the discussion and investigation of the different types of strategy for different types of problem. Each task has a grid showing which strategies are suggested for the various elements within that task. For example, the grid on the right shows which strategies children are likely to use in their investigation of the **Balancing Act** task. These strategies are exemplified in the **Connect** stage of the task.

Draw a diagram	✓
Make a table / spot a pattern	
Make a model / act it out	✓
Try then improve	
Create a systematic list	
Solve a simpler but related problem	✓
Work backwards	

The route through a task

The tasks are designed to allow a degree of flexibility, according to the confidence of the children and the time available. The **Teacher Guide** offers a route through each task, using the 6Cs. Each task could take a part of a lesson, a full lesson or several lessons. Children are given extensive opportunities to explore mathematical concepts and to discuss and share ideas. Some children will make links more quickly and will be able to skip some of the steps.

Full and Express Pathways

There are two main ways to use each task: the **Full 6C Pathway** and the **Express Pathway**. The Full Pathway exemplifies each stage of the journey through the 6Cs. However, there will be some occasions where there may not be sufficient opportunity to complete the Full Pathway, and teachers may choose to use the Express Pathway instead.

A Full Pathway could take a full lesson or several lessons, whereas an Express Pathway could form part of a lesson or could be returned to over a series of lessons.

Each task has two **Pupil Resource** pages for the **Curious** task: the first page shows the 'Full' version, and the second page shows a slightly modified 'Express' version. When following the **Full Pathway**, teachers should use the prompts in the Awe and wonder and Develop and deepen sections of the **Teacher Guide**, to encourage discussion and exploration, and allow children to spend a significant amount of time developing their own mathematical questions.

Although the **Express Pathway** prompt is similar, it is a little more direct and focused. A much shorter discussion would take place, before moving fairly quickly onto the **Connect** tasks. Teachers can then choose how much of the rest of the cycle the children engage in. Although the Express Pathway will involve fewer stages, the quality of the explanation or response should remain high. The expectation is that children would still explore the structure of the mathematics rather than just replicate the examples provided in an attempt to provide proof.

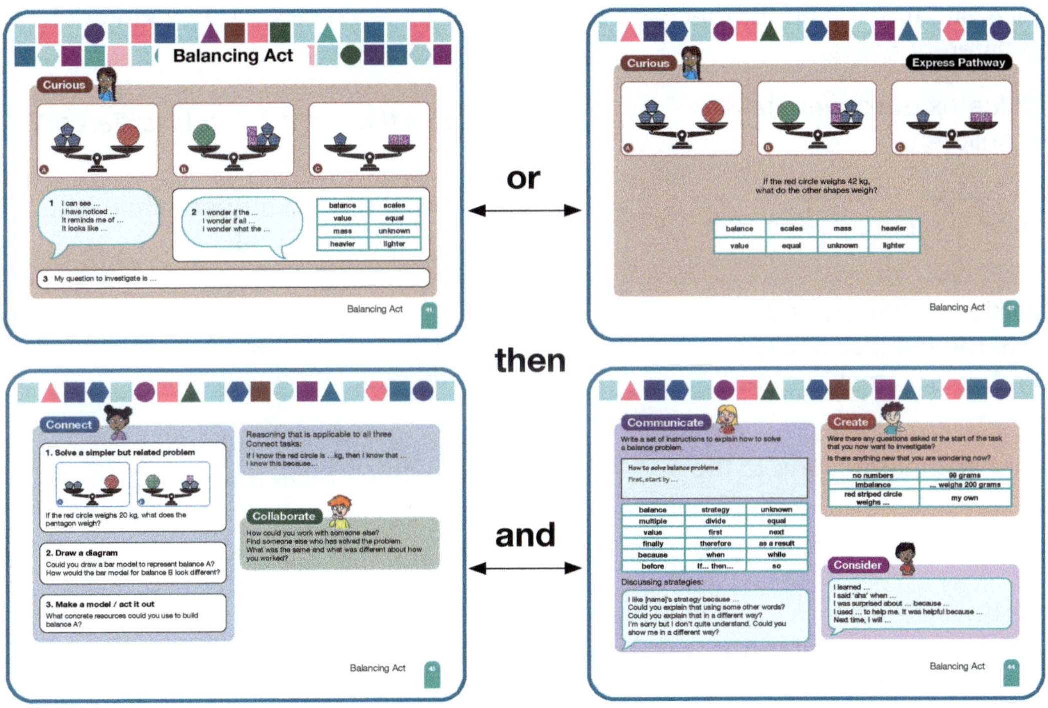

Each Pathway has its own **Curious Pupil Resource** page**.**

The tasks and the **Pupil Resource** pages for the remaining 5Cs are common to both pathways.

Stumble support

The in-built 'stumble support' provides inclusive greater-depth opportunities for all by enabling scaffold-enabled access. You will find explicit references to 'stumble support' within the notes section of some of the **Teacher's Presentation** slides, but stumble support is also woven throughout all the components, in the form of word banks, prompts and sentence stems.

The **How to use** sections below explain the different 6C elements within each task, briefly describing the purpose of each element, key planning questions and details of any additional scaffolds provided. It is useful to look at the **Teacher Guide** and the **Pupil Resources** together, so that decisions concerning when to distribute the **Pupil Resources** can be made. The **Pupil Resources** do not have to be shared immediately and in some cases, the **Curious** stage will give guidance on how to introduce the question / problem to the children *before* the **Pupil Resources** are shared.

The **Teacher's Presentation** slides provide flexibility in terms of display and also ongoing 'stumble support'.

How to use the Teacher Guide

The tasks work with a series of prompts from the teacher, that can be used as whole-class prompts, or with smaller groups. These include key structural questions and prompts such as:

- *What's the same? What's different?*
- *Which is harder / which is easier?*
- What happens if I change ... ?
- If I know this, I also know ...

The prompts aim to elicit understanding of concepts, to get children thinking about how they can apply ideas to unfamiliar situations, and for teachers to assess whether children are making rich connections between different areas.

Italic bulleted text indicates questions the teacher could ask the children. →

- *What's the same? What's different?*

Other bulleted text indicates exemplar statements, questions or sentence stems from children. →

- If I know the red striped circle is ... kg, then I know that ...

Where the first few words are **underlined**, this indicates the words the teacher could use to prompt children towards making the observation. →

- <u>I wonder</u> if the shapes have the same value in each picture.

Non-bulleted text indicates instructions to the teacher. →

When children have developed their questions, reveal that the red striped circle weighs 42 kg.

A **yellow warning symbol** indicates important things to watch out for: common misconceptions that need addressing in case they lead children down a false path, or things that must be done before moving on to the next part of a task. →

⚠ Watch out for children who think that the circles have to have the same value in each diagram.

The **Teacher Guide** tends to follow the same structure for each task. Below is an explanation of each section, using the task **Balancing Act** for illustrative purposes.

Text in blue below indicates questions teachers may want to ask themselves when they get to each section of a task.

Introduction

This explains the aims of the task and how it delivers greater depth.

Balancing Act

Year 3

Introduction

Balancing Act uses images of different shapes on balance scales. Children use the mass of one shape to work out the masses of other shapes on other balance scales.

Children investigate similar problems, working with others to share ideas, and they create their own problems, all the while communicating their ideas and findings.

This task promotes greater-depth learning in a range of ways, supporting children to identify and articulate underlying mathematical structures. For example, in the Curious stage, greater depth is promoted by asking children to make rich mathematical contributions to discussions and suggest a variety of appropriate mathematical questions.

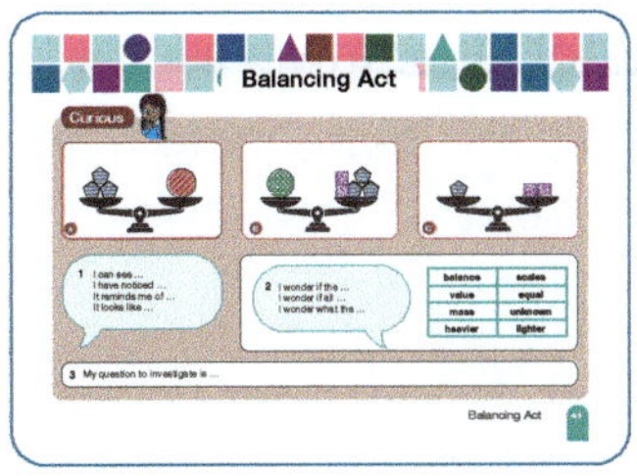

This **Teacher Guide** is supported by photocopiable **Pupil Resources** and **Teacher's Presentation** PowerPoint slides for whole-class or group display.

National Curriculum

All of the Lower KS2 tasks can be done in either Year 3 or Year 4, but a particular Year group is suggested for each task.

Consider children's prior learning and experiences for each of the noted National Curriculum statements when deciding when to use each task.

Do children have sufficient skills and knowledge to access this task?

National Curriculum

Balancing Act covers these Year 3 National Curriculum objectives.

Measurement

- *The comparison of measures includes simple scaling by integers (for example, a given quantity or measure is twice as long or five times as high) and this connects to multiplication.* [non-statutory guidance]

Multiplication and division

- Solve problems, including missing number problems, involving multiplication and division, including positive integer scaling problems and correspondence problems in which *n* objects are connected to *m* objects.

The 6C cycle

The relative weightings of the 6C elements are given for each task to help teachers assess which tasks are suitable for their class at different times.

The 6C cycle

The grid to the right shows the relative weighting given to each of the 6 Cs in **Balancing Act**.

Curious	☆☆☆
Connect	☆☆
Collaborate	☆
Communicate	☆☆
Create	☆☆
Consider	☆☆

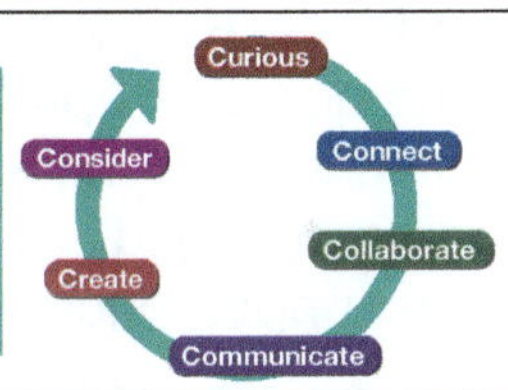

Problem-solving heuristics

For each task, specific problem-solving strategies are recommended for children to use during their investigation.

These strategies are then exemplified in the **Connect** section of the **Teacher Guide.**

Which strategies are children already aware of?

Is this an opportunity to teach a new strategy or apply a pre-existing skill?

Which strategy will be the focus of your modelling?

Problem-solving heuristics

The grid shows which problem-solving strategies children are likely to use when investigating the Connect tasks in **Balancing Act**. These strategies are further exemplified in the Connect stage below (pages 121–123).

Draw a diagram	✓
Make a table / spot a pattern	
Make a model / act it out	✓
Try then improve	
Create a systematic list	
Solve a simpler but related problem	✓
Work backwards	

Key mathematical ideas

Notes are provided to support teachers' subject knowledge and outline the possible pedagogical pathways.

Are there any aspects you need to clarify before the teaching begins?

Key mathematical ideas

Balancing Act is a task that builds secure foundations for later work on algebra in Year 6 and in Key Stage 3. This task does not need children to record algebraically, but it provides a conceptual opportunity to experience that using letters algebraically is simply short-hand for representing a model which can be drawn or built at this stage.

Children are presented with several balance scales with different shapes on each side of the balance. The masses of the shapes are not given. Children are encouraged to look at each balance scale in turn and describe what they see, and then to compare the different scales, focusing on what is the same and what is different.

Children might say or write statements such as:

- 3 pentagons are equal to 1 red striped circle. / 3 pentagons = 1 red striped circle (balance A)
- The green spotty circle is the same as 2 squares and 3 pentagons. / 1 green spotty circle = 2 squares + 3 pentagons (balance B)
- 1 pentagon weighs the same as 2 squares. / 1 pentagon = 2 squares (balance C)

They will develop a range of statements that describe relationships between the shapes. For example, 'the mass of the green spotty circle is the same as 2 squares and 1 red striped circle'.

When working at greater depth, children look for shortcuts and might write 'p' instead of 'pentagon' or 'r' instead of 'red circle', so look out for representations such as '$3p = 1r$'.

The mass of the red striped circle (42 kg) should be shared once the children have completed **Develop and deepen**. Once they know the mass of one shape, they can work out the mass of other shapes.

Using the initial statement '3 pentagons are equal to 1 red striped circle', children might now say: 'I can divide 42 kg by 3 to find the mass of a pentagon because the mass of 3 pentagons is 42 kg.'

3 pentagons = 42, so 1 pentagon = $42 \div 3 = 14$ kg

They can then use this value to work out the mass of the other shapes:

1 pentagon = 2 squares (from slide 3)

14 = 2 squares, so 1 square = $14 \div 2 = 7$ kg

and

1 green spotty circle = 2 squares + 3 pentagons (from balance B)

1 green spotty circle = $2 \times 7 + 3 \times 14 = 14 + 42 = 56$ kg.

Answers to the tasks are given on page 125.

Curious

For the Full 6C Pathway, advice is provided for teachers on how to introduce the question / problem to maximise curiosity.

Two progressive elements can be used:

- **Awe and wonder;**
- **Develop and deepen.**

Awe and wonder gets children talking about the question / problem and allows them to share their initial thoughts and start to become engaged.

'Stumble support' includes:

- examples of possible responses from children, with underlined words indicating how the teacher might prompt children towards these observations.

Curious

Awe and wonder

Reveal slides 1–3 one at a time, allowing time for discussion and observations as prompted below.

Slide 1

- I can see three pentagons and a circle.
- I have noticed that the circle is larger than a pentagon.
- It reminds me of a seesaw in the park.
- It looks like the shapes are equal in mass if it is a balancing scale.
- 3 pentagons are the same / have the same mass as a circle.
- It reminds me of a number sentence: 3 pentagons = 1 red circle.

⚠ Watch out for children who use 'weight' instead of 'mass'. Weight/mass may be used interchangeably in KS1. In KS2 it is important that children use mass (kg/g).

Slide 2

- I can now see a circle, 2 squares and the 3 pentagons.
- I have noticed that the number of pentagons is the same in both pictures.
- I have noticed that the circles are different colours.

Slide 3

- I have noticed that the pentagon appears again.
- I can see that 2 squares are equal to a pentagon.

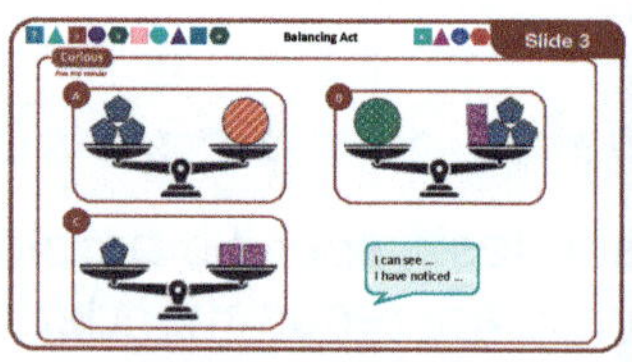

Develop and deepen aims to help children develop more mathematically accurate lines of enquiry, and to funnel children towards the pre-identified questions.
'Stumble support' includes:

- word banks to lift mathematical vocabulary and help direct children to appropriate lines of enquiry;
- examples of age-appropriate lines of enquiry.

How will the word bank be shared with children?

How will the possible lines of enquiry be collected? Whole class, small groups and so on. Do they need to be visible and shared with all to encourage children building on each other's ideas?

Will children all work on one line of enquiry or will they be able to select their preferred line of enquiry?

Curious

Develop and deepen

Slide 4

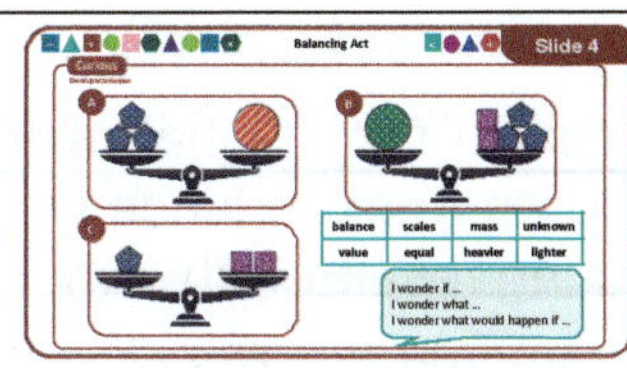

Display slide 4. Encourage children to use the words in the word bank to develop and deepen their thoughts and questions.

Possible lines of enquiry

- I wonder if all of the shapes represent a different mass.
- I wonder what the mass of the red striped circle is.

balance	scales
value	equal
mass	unknown
heavier	lighter

Possible lines of enquiry linked to the word bank:

- I wonder if the shapes have the same value in each picture.
- I wonder if the red striped circle is lighter than the green spotty circle as there are 2 squares with the 3 pentagons in the second balance.

Possible lines of enquiry that children may raise now, but should be held back until the **Create** task:

- I wonder what would happen to the values if one of the balances wasn't equal.

Consider whether / how to collate and display all the suggested questions, and whether / how to steer children towards selecting an appropriately challenging question.

Connect

The aim of the **Connect** task is to support children in using a range of problem-solving strategies, to try out different 'ways in' to the problem.

Guidance is given on the strategies best suited to the task and how they work in the given context.

Most **Connect** tasks will deliver a number of suitable alternative strategies. Children could work through all of the tasks suggested, or choose just one or two.

'Stumble support' includes:

- additional prompts, questions and suggestions for teachers to help children be successful with their chosen strategies.

Will children work through all of the Connect tasks, or just one or two?

What resources will need to be provided?

How will the 'stumble support' be made accessible?

Connect

Three tasks involving different strategies are suggested for children to make connections.

1. Solve a simpler but related problem

Slide 9

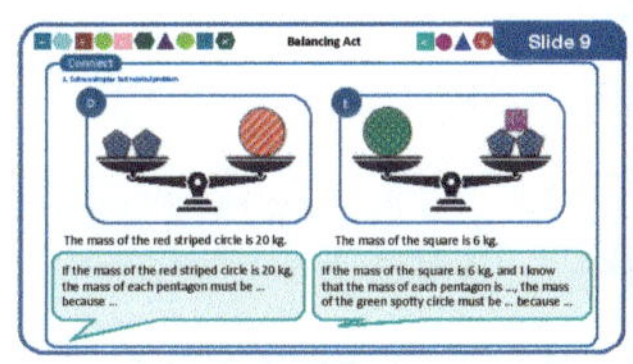

Display slide 9 (balances D and E).

Focus children's attention on balance D.

- *If the mass of the red striped circle is 20 kg, what is the mass of the pentagon?*
- If I know the red striped circle is … kg, then I know that … because
- *Can you draw a diagram to prove your answers?*

Now, look at balance E.

- *What is different this time?* (From balance D, now we know the masses of the pentagons.)

⚠ Watch out for children who think that the circles have to have the same value in each diagram.

2. Draw a diagram

Bar models provide a clear visual aid to allow children to compare the size of each shape and see how to find an unknown mass or value. For example, for balance A, if the mass of the red striped circle is known, then to find a pentagon, this needs to be divided by 3. As more unknowns are discovered, encourage the children to annotate their diagrams. Once the mass of the pentagon is known, this unlocks the mass of all of the other shapes.

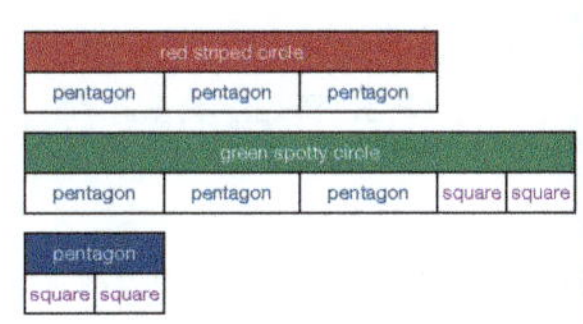

- *Have you shown that all the pentagons are the same value in your bar models? Make sure they are all the same size.*
- *If we know the value of the red striped circle, how can we find the value of a pentagon?*
- *Where else can we put this information on our diagrams?*
- *If we know this, what else do we now know?*

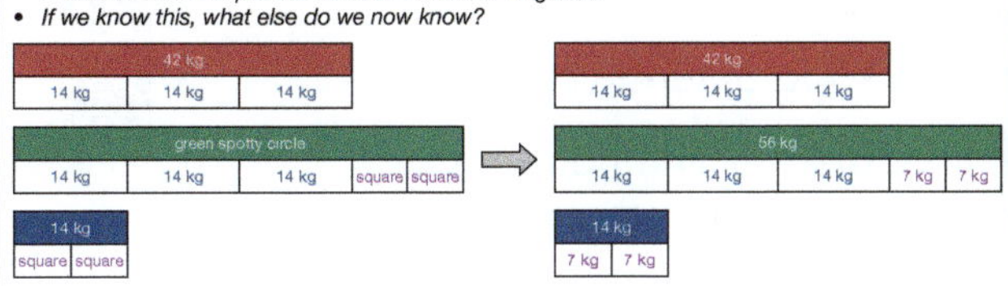

Collaborate

The aim of the **Collaborate** task is for children to experience working with, or sharing results with, other children.

Various models of collaboration are suggested.

Which group-work skills will be a focus for your children?

Collaborate

Refer to the Collaborate task of the **Pupil Resource**.

Pairs of children who worked on the same task in the Connect stage could compare how they solved the problem.

- *What was the same? What was different?*

Communicate

The aim of the **Communicate** task is to get children to think about what they have discovered and how they are going to share their findings with others.

There may be opportunities for cross-curricular links to the National Curriculum.

'Stumble support' varies by task but may include:

- writing frames;
- word banks to deepen understanding;
- sentence stems.

How will respectful talk be modelled?

How and where will this be recorded?

Communicate

Refer to the Communicate task of the **Pupil Resource**.

This task provides an opportunity for the children to write instructions.

How to solve balance problems
First, start by ...

Children can use the word bank for vocabulary support.

balance	**strategy**	**unknown**	**multiple**	**because**	**when**
divide	**equal**	**value**	**first**	**while**	**before**
next	**finally**	**therefore**	**as a result**	**if ... then ...**	**so**

Afterwards, encourage children to compare the strategies that they have used, and ask each other questions for clarification.

- I like [name]'s strategy because ...
- Could you explain that in some other words?
- Could you explain that in a different way?
- I'm sorry but I don't quite understand. Could you show me in a different way?

Create

The aim of the **Create** task is to inspire children to continue with developing their own questions to further deepen learning.

Examples of further lines of enquiry are provided in the **Teacher Guide** for each task.

Unanswered questions from the **Curious** stage can be returned to and engaged with.

'Stumble support' includes:

- sentence stems;
- word banks to extend vocabulary.

Were any additional lines of enquiry suggested in the Curious task that could be pursued now?

Create

Refer to the Create task of the **Pupil Resource**.

- *Were there any questions asked at the start of the task that you now want to investigate?*
- *Is there anything new that you are wondering now?*

Children can use the word bank to prompt further questions of their own.

no numbers	**99 grams**
imbalance	**... weighs 200 grams**
red striped circle weighs ...	**my own ...**

- I wonder what would happen to the values if one of the balances wasn't equal.
- I wonder if it would work if no numbers were involved.
- I wonder what would happen if the red striped circle wasn't 42 kg but was 99 grams.
- I wonder if I could start with a different shape. For example, what if the square weighed 200 grams?
- I wonder if I can design my own puzzle using these shapes and balances.

Slide 15

If needed, display slide 15.

- *What's the same and what's different in these diagrams compared to balances A–C?*
- This time one of the balances is not equal.
- The two squares must be lighter than a pentagon rather than equal to it.

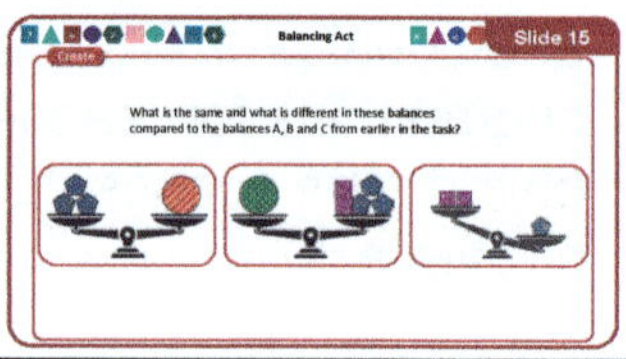

Consider

The aim of the **Consider** task is to get children to reflect on what they have learned.

- Cognitive domain – what have they learned?
- Affective domain – how do they feel about the learning?
- Metacognitive domain – how did they learn it?

'Stumble support' includes:

- additional sentence stems.

Which of the suggested sentence stems will you encourage children to consider and complete?

Consider

Refer to the Consider task of the **Pupil Resource**.

Encourage children to think about the strategies that they have used and also the instruction writing of the other children.

The prompts below may help get children started.

Cognitive learning domain

- The trickiest part of solving my question was …
- I learned …
- The best strategy was …

Affective domain

- The fun bit of my learning was … It was fun because …
- I said 'aha' when … I said it because …
- I was surprised about … because …

Metacognitive domain

- I got stuck when … I unstuck myself by …
- I used … to help me. It was helpful because …
- Next time, I will …

Answers

Where appropriate, worked solutions and answers are provided.

For some tasks, these will be presented on the final page of the **Teacher Guide** for that task. For others, answers are given in situ throughout, because of the cumulative nature of the tasks.

Answers

The three balance scale images can be represented like this:
3 pentagons = 1 red striped circle
1 green spotty circle = 2 squares + 3 pentagons
1 pentagon = 2 squares

Substituting the mass given for the red striped circle, 42 kg, gives:
[using balance A] 3 pentagons = 42 kg, so 1 pentagon = 42 kg ÷ 3 = 14 kg

Therefore:
[using balance C] 14 kg = 2 squares, so 1 square = 14 kg ÷ 2 = 7 kg

And so:
[using balance B] 1 green spotty circle = 2 × 7 kg + 3 × 14 kg = 14 kg + 42 kg = 56 kg

Connect: **Solve a simpler but related problem**

The balances on slide 9 can be represented as:
2 pentagons = 20 kg
1 green spotty circle = 2 pentagons + 6 kg

Working out the mass of 1 pentagon:
2 pentagons = 20 kg, so 1 pentagon = 20 kg ÷ 2 = 10 kg

And therefore:
1 green spotty circle = 2 × 10 kg + 6 kg = 26 kg

Create

The unbalanced scale in slide 15 can be represented using the inequality sign:
1 pentagon > 2 squares

How to use the Pupil Resources

Full Pathway Curious page

Curious question / problem
This acts as an *aide memoire* for the rest of the cycle.

Awe and wonder
Sentence stems to support children to articulate their initial observations and ideas.

Chosen line of enquiry
A prompt for children to select a line of enquiry they would like to pursue.

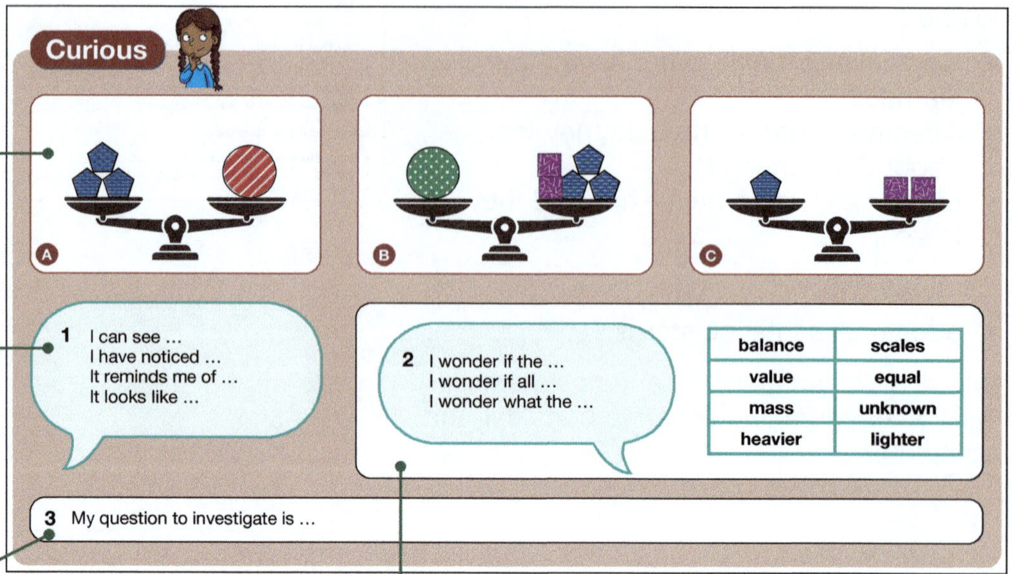

Develop and deepen
A word bank to boost mathematical vocabulary and support mathematically accurate questions.
Sentence stems to support children to articulate their questions.

Express Pathway Curious page

Curious question / problem
This acts as an *aide memoire* for the rest of the cycle.
The question / problem is presented in a more focused way.

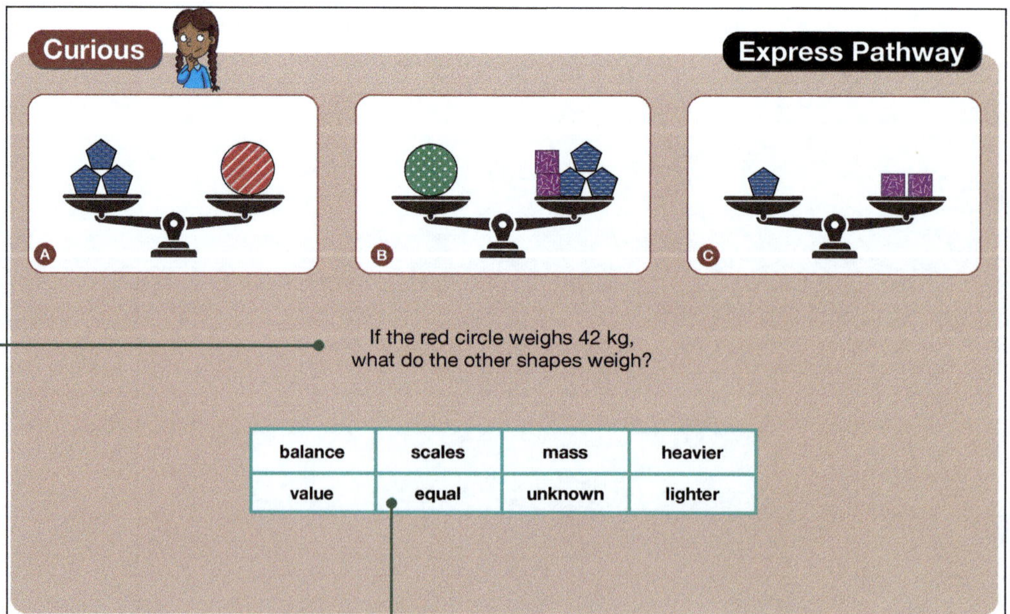

A word bank to boost mathematical vocabulary.

The remaining 5 Cs, common to both pathways

Connect
Children are presented with a range of strategies. They could select their preferred option, or try out more than one.

Connect

1. Solve a simpler but related problem

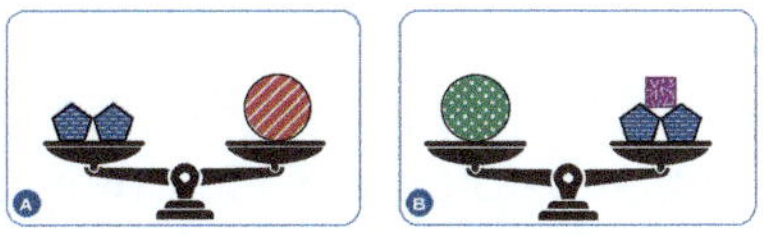

If the red circle weighs 20 kg, what does the pentagon weigh?

2. Draw a diagram

Could you draw a bar model to represent balance A?
How would the bar model for balance B look different?

3. Make a model / act it out

What concrete resources could you use to build balance A?

Reasoning that is applicable to all three Connect tasks:

If I know the red circle is …kg, then I know that …
I know this because…

Collaborate

How could you work with someone else?
Find someone else who has solved the problem.
What was the same and what was different about how you worked?

Collaborate
Children are prompted to work with others, and consider helpful ways of working.

Create
Children are prompted to continue developing their own questions to further deepen their learning. A word bank encourages mathematically accurate language.

Communicate
Children are asked to explain their findings clearly, and are given suggestions for how their ideas could be shared.
Word banks, sentence stems and writing frames are provided where appropriate.

Communicate

Write a set of instructions to explain how to solve a balance problem.

How to solve balance problems
First, start by …

balance	strategy	unknown
multiple	divide	equal
value	first	next
finally	therefore	as a result
because	when	while
before	If… then…	so

Discussing strategies:

I like [name]'s strategy because …
Could you explain that using some other words?
Could you explain that in a different way?
I'm sorry but I don't quite understand. Could you show me in a different way?

Create

Were there any questions asked at the start of the task that you now want to investigate?

Is there anything new that you are wondering now?

no numbers	99 grams
imbalance	… weighs 200 grams
red striped circle weighs …	my own

Consider

I learned …
I said 'aha' when …
I was surprised about … because …
I used … to help me. It was helpful because …
Next time, I will …

Consider
Sentence stems guide children in reflecting on what they have learned, how they learned it and how they feel about their learning.

Assessment rubric

The assessment rubric has been designed to identify the stepping stones of attainment across the 6Cs. For a child to be described as working at greater depth, they need to be performing highly in all components. Most children working at greater depth will have a mixed profile between consolidating and extending. The rubric should be used with the whole of the class.

Using a rubric can help with:

- developing self-regulation: children can use it to help plan or monitor their own learning;
- direct teaching: it moves the focus from task completion to learning skills, knowledge and behaviour;
- providing a learning pathway: when used over time, it can help to identify gaps and next steps so feedback can be fine-tuned and highlighted as a teaching opportunity; this feedback and teaching then loops back into the next task.

Lower Key Stage 2 assessment rubric

	Curious	
	Awe and wonder	*Develop and deepen*
Not evident		
Entering	Makes limited contributions. Contributions made are not connected to the task.	Makes unconnected statements / questions when prompted.
Beginning	Makes simple contributions when prompted, e.g. recalls key fact or simple pattern.	Contributes statements rather than questions with limited mathematical content, sometimes needs prompting.
Developing	Contributes to discussion, e.g. notices a common property or pattern.	Contributes statements rather than questions.
Consolidating	Contributes to discussion, identifies connections and patterns.	Offers a single or limited number of mathematical questions.
Extending	Contributes to discussion. Notices and queries connections and patterns.	Offers a variety of good mathematical questions.

	Connect		Collaborate		
	Application	*Reasoning*	*Individual responsibility*	*Teamwork*	*Respect for others*
Not evident					
Entering	Has trouble figuring out how to solve the problem. Needs heuristic to be chosen for them.	Not sure if their answer is correct or not. Describes what they did.	Often or always relies on others to do the work.	Only shares ideas and suggestions when prompted to.	Impolite or unkind to the group.
Beginning	Needs some help to choose an appropriate heuristic.	With help can figure out if they have a correct answer. Attempts an explanation, but it is incorrect.	Participates but needs to be prompted and encouraged throughout.	Needs occasional prompting to share ideas and suggestions.	Sometimes ignores the feelings of others, e.g. interrupting or ignoring ideas.
Developing	Able to select an appropriate heuristic from a predefined list.	Provides a correct answer but may get confused on the way. Offers an explanation but it is incomplete / unsystematic.	Participates but needs occasional prompting. Gets upset if they do not get to do what they want.	Shares ideas and makes suggestions but gets upset if they are rejected by the group.	Usually polite and kind to the group.
Consolidating	Selects an appropriate heuristic. Can explain why they have chosen it drawing on previous experience.	Provides a correct answer. Explains the property / pattern or rule and how it works. A systematic approach is present, if relevant.	Participates within the group without arguing.	Offers suggestions and ideas. Listens carefully to other group members.	Polite and kind to all group members.
Extending	Selects the most appropriate heuristic and can explain choice including why not other heuristics.	Understands that sometimes more than one answer is possible. Explains the property / pattern or rule and how it works, using examples.	Participates readily and is willing to change roles.	Offers ideas and suggestions. Accepts if they are rejected by the group. Listens carefully and builds on other group members' views.	Polite, kind and encouraging to group members. Disagrees respectfully.

	Communicate		Create
	Clarity	*Mathematical language, images and diagrams*	
Not evident			
Entering	Communication is confusing and difficult to follow.	No mathematical language.	Not able to suggest any further questions.
Beginning	Communication is hard to follow, but can answer questions about learning if prompted.	Attempts to use some mathematical language.	Needs help to develop new relevant questions.
Developing	Communication is hard to follow in places.	Uses some mathematical language supported by a diagram / image / model which is incomplete or slightly incorrect.	Builds on others' questions, changing them slightly.
Consolidating	Communication is clear and easy to follow.	Mathematical language is accurate and supported with correct diagrams / images / models.	Creates their own questions.
Extending	Communication is clear, accurate and well organised.	Mathematical language is consistently accurate and diagrams / images / models are regularly used to illustrate key ideas.	Creates their own questions that are creative and inventive.

	Consider		
	Cognitive	*Affective*	*Metacognition*
Not evident			
Entering	Confuses learning outcomes with title of the task.	Only states that the learning was fun or not, struggles to explain why.	Struggles to identify anything they did when they got stuck. Passively waits for an adult to support them.
Beginning	Attempts to describe what they did, but it is unclear.	Can state that the learning was fun or not; with prompting they can offer a simple reason why. Focuses on their feelings within the group.	Requires prompting to identify which learning behaviours they found helpful when they got stuck and if anyone else helped them.
Developing	Able to identify a simple learning outcome.	Able to identify when things got challenging. Identifies the learning as fun or not, and can explain why. Can describe how they worked with others.	Identifies which learning behaviours they found helpful when they got stuck and, if anyone else helped them, how.
Consolidating	Able to identify what they have learned. Can see links to previous learning.	Able to explain how their feelings changed throughout the task. Describes the learning as engaging or not engaging and offers a reason why. Can describe how they tried to establish working relationships with others.	Explains which learning behaviours were useful and which were not. Includes details on what they could do differently next time.
Extending	Able to identify multiple learning outcomes. Can see links between this learning and previous learning experiences. Can suggest where this learning may be useful in the wider curriculum.	Reflects on how they were feeling throughout the task and why. Able to identify what was and was not engaging about the task; if appropriate they make suggestions to make the task more engaging. Able to identify how they established good working relationships with others and how their learning benefited from interactions with others.	Explains which learning behaviours were useful and which were not. Includes details on how they applied learning from previous tasks and what they could do differently next time.

Who Am I?

Year 3

Introduction

Who Am I? presents a set of clues to a mystery number. It provides the opportunity for children to draw on their understanding of place value, number sense and properties of numbers. In addition, it encourages them to consider the relative usefulness of the clues provided and work systematically to eliminate numbers from their enquiry.

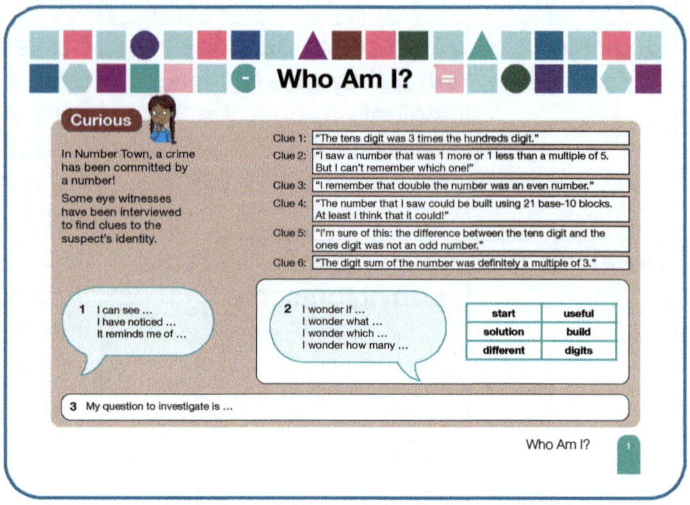

The mystery number is 264. However, this number cannot be found with the clues that are given. When they have worked through all of the clues, children will be left with two numbers: 264 and 399. Children could be tasked to write a final clue themselves, or they could be provided with this additional clue: *The number was 10 less than an even number.*

This task promotes greater-depth learning in a range of ways, supporting children to identify and articulate underlying mathematical structures. For example, in the **Collaborate** stage, greater depth is promoted by asking children to offer suggestions, listen carefully to others and build upon their own ideas, and be able to accept it if their ideas are rejected by the group.

This **Teacher Guide** is supported by photocopiable **Pupil Resources** and **Teacher's Presentation** PowerPoint slides for whole-class or group display.

National Curriculum

Who Am I? covers these Year 3 National Curriculum objectives.

Number and place value

- Solve number problems and practical problems involving these ideas.*

*See National Curriculum statutory requirements for the full list of statements to which this refers.

The 6C cycle

The grid to the right shows the relative weighting given to each of the 6 Cs in **Who Am I?**

Curious	★★
Connect	★★
Collaborate	★★★
Communicate	★★★
Create	★★
Consider	★★

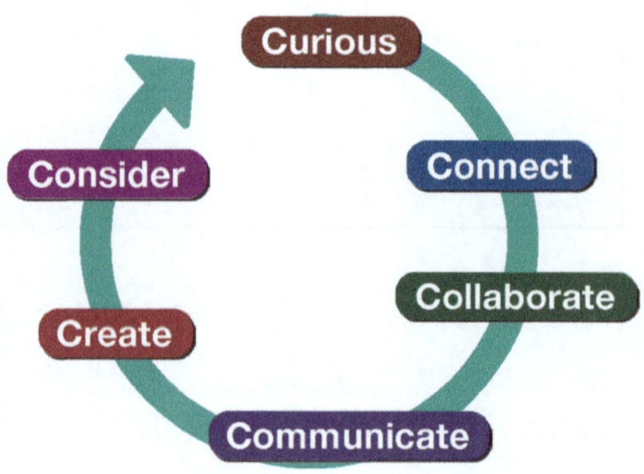

Problem-solving heuristics

The grid shows which problem-solving strategies children are likely to use when investigating the **Connect** tasks in **Who Am I?** These strategies are further exemplified in the **Connect** stage below (pages 28–29).

Draw a diagram	
Make a table / spot a pattern	
Make a model / act it out	✓
Try then improve	✓
Create a systematic list	✓
Solve a simpler but related problem	
Work backwards	✓

Key mathematical ideas

Children will need to consider a sensible order in which to investigate the clues, work methodically and keep track of the possibilities. They should aim to start with a clue that narrows the options down as much as possible. For example, the clue 'The tens digit was 3 times the hundreds digit' will allow them to identify three possible pairs of hundreds and tens digits: 13_, 26_ and 39_.

One clue is ambiguous: 'I saw a number that was 1 more or 1 less than a multiple of 5. But I can't remember which one!' Children will need to use additional clues to test whether the possible numbers satisfy both. For example, depending on when children test this clue, they could be left with the numbers 261, 264, 396 and 399. However, neither 261 nor 296 satisfy the clue: 'the difference between the tens digit and the ones digit was not an odd number.'

The tens and ones digits in both 264 and 399 do have a difference that is not odd. This may lead to an interesting discussion over whether zero is an even number.

- *Zero sits between two odd numbers, so I think it is even.*
- *Even + even = even, and 2 + 0 = 2, so zero could be even.*
- *Even ÷ 2 = an integer. We can't divide 0 by 2 and get an integer, so zero can't be even.*

When children test whether 264 can be built using 21 base-10 blocks, they are likely to first try 2 hundreds, 6 tens and 4 ones (12 blocks). They will find that exchanging 1 ten for 10 ones will increase the number of blocks by 9: 2 hundreds, 5 tens and 14 ones (21 blocks).

Children should disregard the clue 'double the number was an even number', recognising that the clue is not helpful because any number doubled will result in an even number.

Although it can be inferred that the mystery number is a three-digit number, some children may like to consider the possibility of it being a four-digit number and how this would affect numbers generated using the clues. For example, 'The digit sum of the number was definitely a multiple of 3' would affect the outcome, but 'I saw a number that was 1 more or 1 less than a multiple of 5. But I can't remember which one!' would not.

Two possible routes through the clues to identify the final two possibilities are shown on pages 32–33.

Teaching guidance

If you are using the **Express Pathway**, please turn to pages 27–31.

Teaching guidance for the **Full 6C Pathway** is given below.

The **Teacher's Presentation** PowerPoint file has slides specifically mentioned in this teaching guidance. Some of the slides include notes (within the PowerPoint file) providing 'stumble support' suggestions, that may help you to support children in their problem-solving thinking and enable access for children who are less confident.

Awe and wonder

Slide 1

Display slide 1.

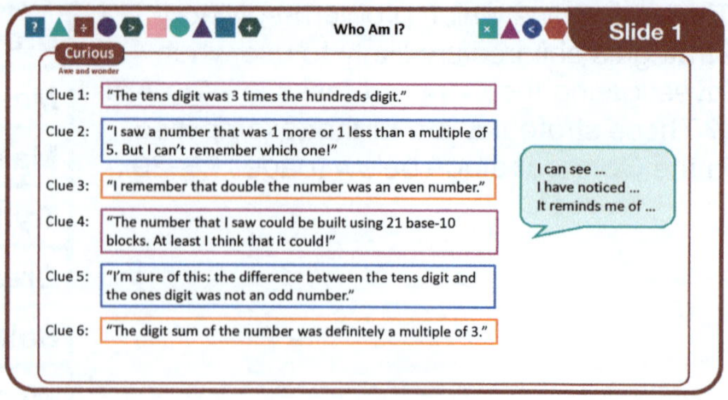

- I can see that there are six clues.
- I have noticed that 'even' is mentioned in two of the clues (but one of the clues says 'not odd', rather than 'even').
- I have noticed that it must be at least a three-digit number because hundreds, tens and ones are mentioned.
- I have noticed that some of the witnesses are unsure about what they saw.
- It reminds me of trying to work out a riddle.
- I can see that one of the clues says 'was not odd' so it must be the opposite.

Develop and deepen

Slide 2

Display slide 2.

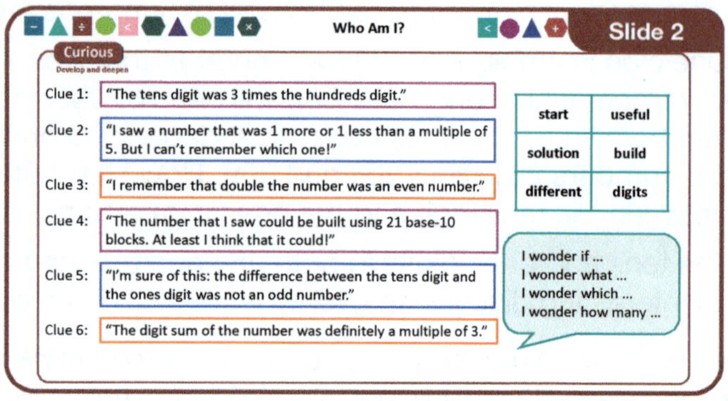

Encourage children to use the words in the word bank to develop and deepen their thoughts and questions.

start	**useful**	**solution**
build	**different**	**digits**

Possible lines of enquiry:

- I wonder if one of the digits is 3.
- I wonder how many digits are in the number.
- I wonder if it is a three-digit number, as thousands are not mentioned.

Possible lines of enquiry linked to the word bank:

- I wonder which clue we should start with.
- I wonder if there is only one solution.
- I wonder if any of the clues are not useful.
- I wonder if you could build the number using a different amount of base-10 blocks.

Possible lines of enquiry that children may raise now, but should be held back until the **Create** task:

- I wonder what would change if it was a four-digit number, but the clues stayed the same.
- I wonder how many different ways you could build the number using base-10 blocks.
- I wonder if I could write my own clues to a mystery number.

Consider whether / how to collate and display all the suggested questions, and whether / how to steer children towards selecting an appropriately challenging question.

Pupil Resources for Full and Express Pathways

Distribute the **Pupil Resources** for **Who Am I?** to the children to help support them in their exploration of the different tasks. Use either of the **Curious Pupil Resources** depending on whether you are following the **Full Pathway** or the **Express Pathway**. The **Pupil Resources** for the remaining 5Cs are common to both pathways.

To provide additional flexibility in the classroom, each of the **Pupil Resources** is also provided in the **Teacher's Presentation** PowerPoint file (see slides 3–6), so you can display these at any time if you wish.

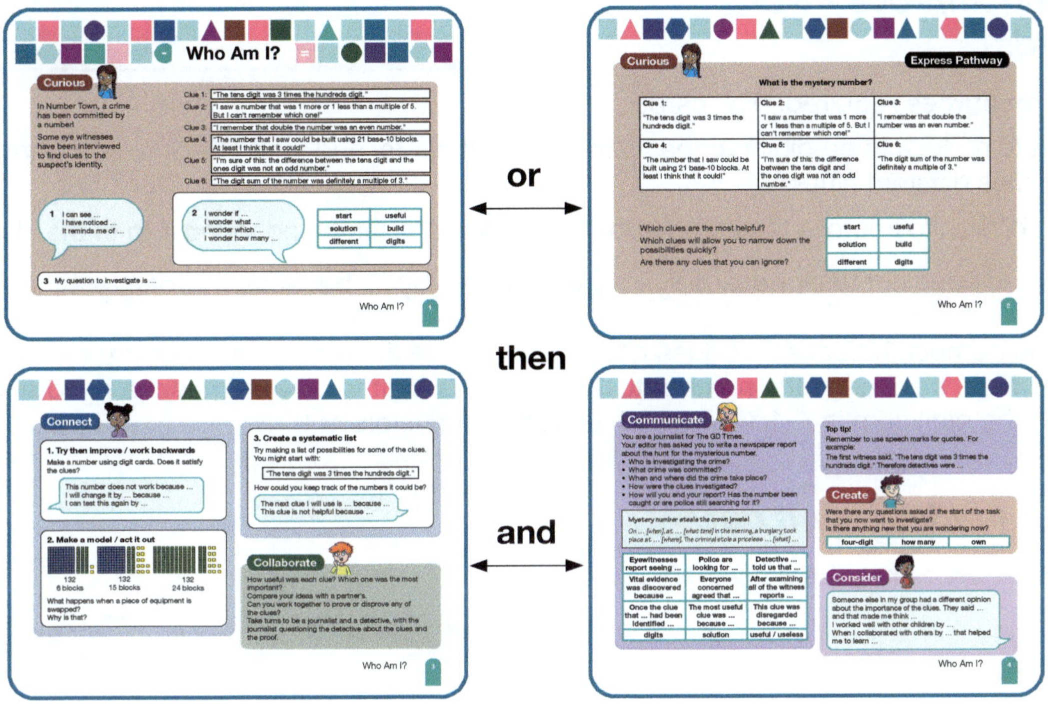

Express Pathway

If you are following the **Express Pathway**, spend a few minutes discussing the **Curious Pupil Resource** with the children. (This is slide 4, so you could display it if you wish.) Use some of the prompts on the previous page to help.

Three tasks involving different strategies are suggested for children to make connections.

Slide 7

Print slide 7. Distribute it to children and ask them to cut out the clues. This will allow children to focus on individual clues, and organise them into a sensible order, throughout the **Connect** tasks.

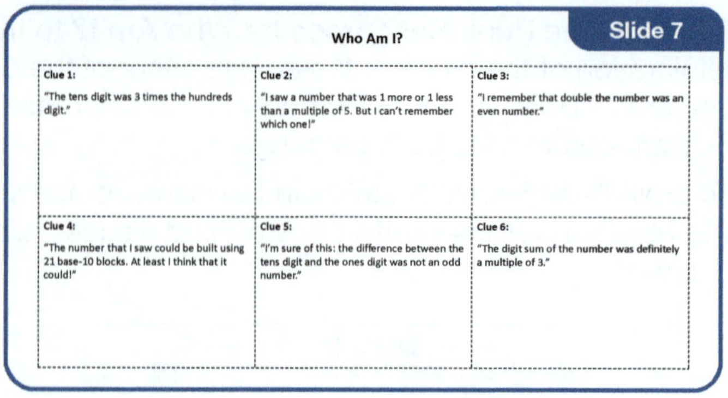

1. Try then improve / work backwards

Additional resources required: digit cards

Ask children to make a three-digit number with their digit cards, and then check to see whether their number satisfies different clues.

Encourage children to improve the number that they have chosen by adapting it to fit the clues.

- This number does not work because … I will change it by … because … I can test this again by …
- I made the number 123, but that does not work because the tens digit is not 3 times the hundreds digit. If I change the number to 133, the tens digit is now 3 times the hundreds digit. However, the digit sum is 7 (1 + 3 + 3), which is not a multiple of 3.

2. Make a model / act it out

Additional resources required: base-10 blocks

Slide 8

Display slide 8.

Ask children to focus on a single clue: 'The number that I saw could be built using 21 base-10 blocks. At least I think that it could!'

Encourage children to build their chosen number using base-10 blocks.

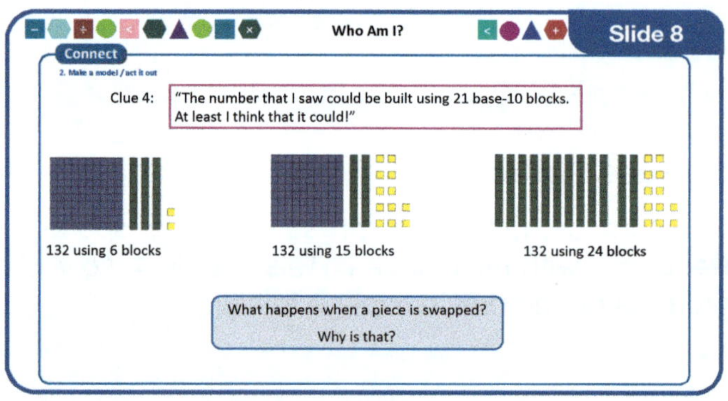

- *How many blocks have you used?*
- *Can you adapt your model to increase or decrease the number of blocks that you have used?*
- *Do you notice anything each time you swap some equipment? Why does this happen?*

Draw out that each time they swap a ten for 10 ones, or a hundred for 10 tens, the number of blocks increases by 9, because 1 block has been removed and replaced by 10 blocks.

⚠ Make sure children are confident at regrouping the correct amount. Check how they count these.

3. Create a systematic list

Additional resources required: digit cards

Slide 9

Display slide 9.

The clue 'The tens digit was 3 times the hundreds digit' is a good place to start to begin to narrow down the possibilities.

- *Which hundreds and tens digits are possible?*
- *How could the 3 times table help?*

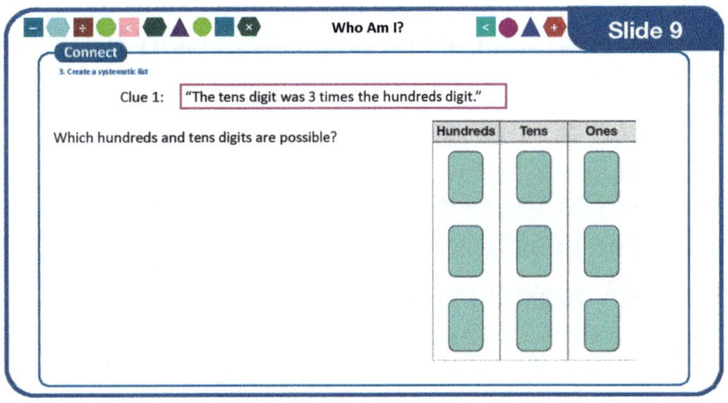

On the slide, write the three possible pairs of hundreds and tens digits on the cards.

Hundreds	Tens	Ones
1	3	
2	6	
3	9	

This initial list can then be built upon by considering another clue, for example, 'I saw a number that was 1 more or 1 less than a multiple of 5. But I can't remember which one!'

- *How will you combine this information with your findings from the first clue?*
- *How will you keep track of the numbers that you have listed?*

131	261	391
134	264	394
136	266	396
139	269	399

An organised list of remaining possibilities after two clues have been followed.

- *Which clue will you use next?*
- *Why have you chosen that clue and not this clue?*

Encourage children to continue to work through the clues (disregarding the one that is not useful) until they have narrowed down the possibilities as much as they can.

⚠ Watch out for children who are following the clues in the order given, rather than considering the usefulness of the clues. The following sentence stems could be provided:

- The next clue I will use is … because …
- This clue is not helpful because …

Collaborate

Refer to the **Collaborate** task of the **Pupil Resource**.

Slide 10

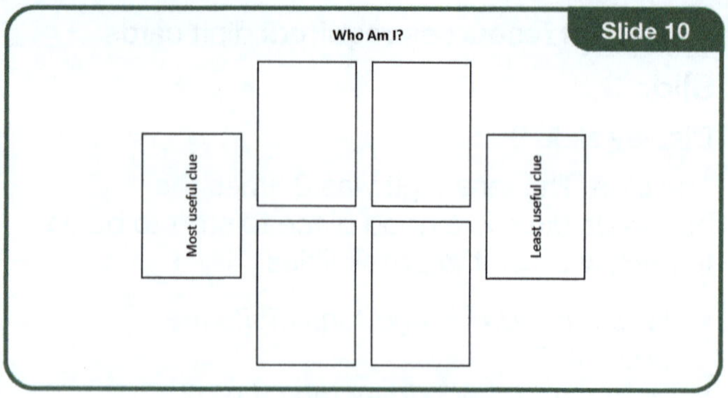

Print slide 10. Distribute it to children so they can position the cut-out clues (from slide 7) in the diamond structure to rank them by usefulness.

Children may like to work on their own to begin with and consider their own starting points and the usefulness of the clues.

Children could be tasked to provide 'proof' that the clue they chose as their starting point is the best option. Children could then meet together as detectives and share their prepared proof.

In pairs, children could take turns as acting as a journalist ready to write their newspaper reports. A detective could be 'put in the hot seat' to find out about the clues and the proof. Children can then swap roles so that they both get the opportunity to ask questions and provide explanations.

- *Which clue allowed you to rule out a lot of suspects early on in the investigation?*
- *How can you prove that this clue means that [a given number] is no longer a suspect?*

Communicate

Refer to the **Communicate** task of the **Pupil Resource**.

Slide 11

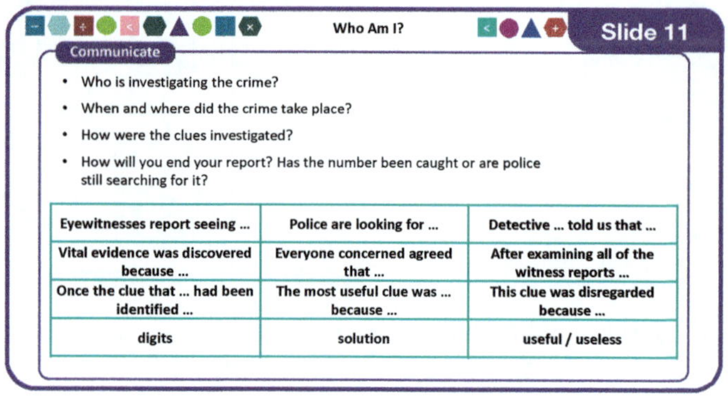

Display slide 11.

Ask children to take on the role of a journalist and write a newspaper article about the hunt for the mystery number. This will give them the opportunity to write animatedly about the problem and how they came to a solution, aiming to create intrigue and suspense for the reader.

Children could consider the following additional details:

- *Who is investigating the crime?*
- *What crime was committed?*
- *When and where did the crime take place?*
- *How were the clues investigated?*
- *How will you end your report? Has the number been caught or are police still searching for it?*

They can use the word bank for vocabulary support.

Eyewitnesses report seeing ...	**Police are looking for ...**	**Detective ... told us that ...**
Vital evidence was discovered because ...	**Everyone concerned agreed that ...**	**After examining all of the witness reports ...**
Once the clue that ... had been identified ...	**The most useful clue was ... because ...**	**This clue was disregarded because ...**
digits	**solution**	**useful / useless**

Refer to the **Create** task of the **Pupil Resource**.

Slide 12

Display slide 12.

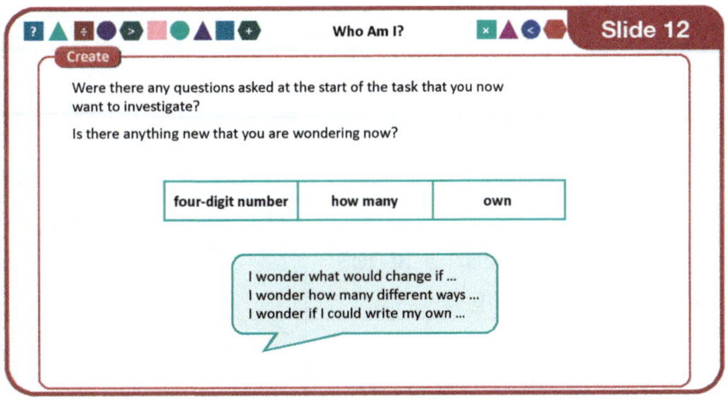

- *Were there any questions asked at the start of the task that you now want to investigate?*
- *Is there anything new that you are wondering now?*

Children can use the word bank to prompt further questions of their own.

four-digit number	**how many**	**own**

- I wonder what would change if it was a four-digit number, but the clues stayed the same.
- I wonder how many different ways you could build the number using base-10 blocks.
- I wonder if I could write my own clues to a mystery number.

Refer to the **Consider** task of the **Pupil Resource**.

Slide 13

Display slide 13.

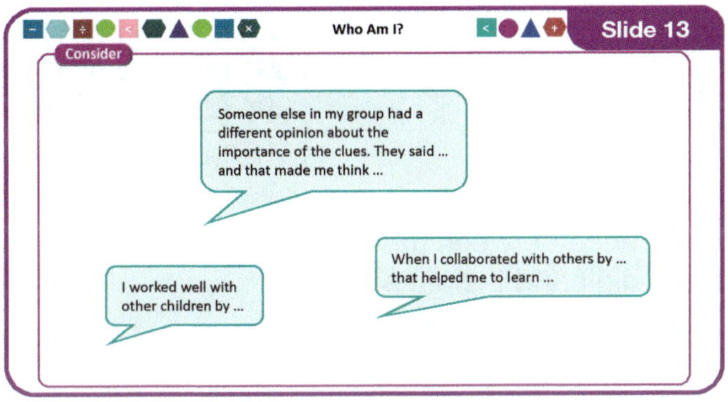

Encourage children to think about the strategies that they have used. The prompts below may help get children started.

Cognitive learning domain

- Someone else in my group had a different opinion about the importance of the clues. They said … and that made me think …
- Someone could argue against my answer by …

Affective domain

- The fun bit of my learning was … It was fun because …
- I worked well with other children by …
- Working as part of a group helped because … It was less helpful when …

Metacognitive domain

- When I collaborated with others by … that helped me to learn …
- Next time I try to solve a problem, I will …

Answers

Here are two possible routes to narrow the solution down to two numbers.

Route 1:

Clue	Reasoning	Possibilities remaining
Clue 1: The tens digit was 3 times the hundreds digit.	The hundreds and tens digits must be either 13_, 26_ or 39_.	13_ 26_ 39_
Clue 2: I saw a number that was 1 more or 1 less than a multiple of 5. But I can't remember which one!	The ones digit must be either 1, 4, 6 or 9.	131 261 391 134 264 394 136 266 396 139 269 399
Clue 6: The digit sum of the number was definitely a multiple of 3.	Numbers eliminated: 131 (digit sum: 5) 134 (digit sum: 8) 136 (digit sum: 10) 139 (digit sum: 13) 266 (digit sum: 14) 269 (digit sum: 17) 391 (digit sum: 13) 394 (digit sum: 16)	261 264 396 399
Clue 5: I'm sure of this: the difference between the tens digit and the ones digit was not an odd number.	The tens and ones digit must either both be even, or both be odd.	264 399
Clue 4: The number that I saw could be built using 21 base-10 blocks. At least I think that it could!	264: 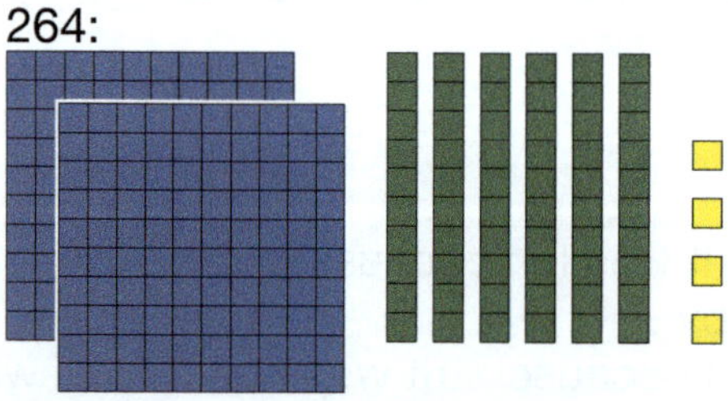 12 blocks Swap a ten for 10 ones (or a hundred for 10 tens) to give 9 more blocks. $12 + 9 = 21$. 399: 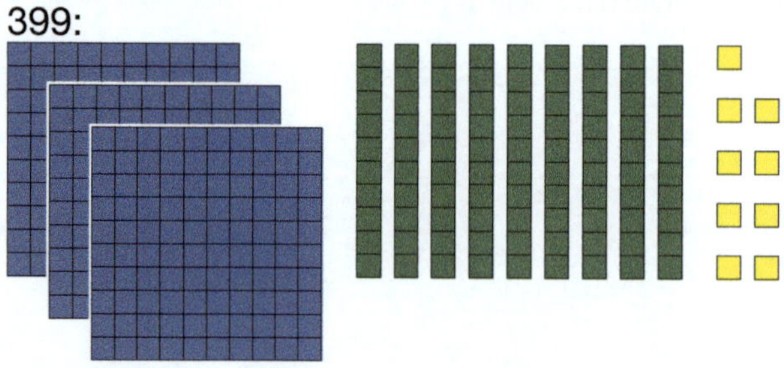 21 blocks	264 399

Route 2:

Clue	Reasoning	Possibilities remaining
Clue 1: The tens digit was 3 times the hundreds digit.	The hundreds and tens digits must be either 13_, 26_ or 39_.	13_ 26_ 39_
Clue 6: The digit sum of the number was definitely a multiple of 3.		132 261 390 135 264 393 138 267 396 399
Clue 5: I'm sure of this: the difference between the tens digit and the ones digit was not an odd number.	The tens and ones digit must either both be even, or both be odd.	135 264 393 399
Clue 2: I saw a number that was 1 more or 1 less than a multiple of 5. But I can't remember which one!	The ones digit must be either 1, 4, 6 or 9.	264 399
Clue 4: The number that I saw could be built using 21 base-10 blocks. At least I think that it could!	264 using 21 blocks: – 1 hundreds-block – 16 tens-rods – 4 ones-cubes 399 using 21 blocks: – 3 hundreds-blocks – 9 tens-rods – 9 ones-cubes	264 399

Place Value Pebbles

Year 4

Introduction

Place Value Pebbles provides an opportunity to explore a traditional number representation (similar to a Chinese abacus) which deepens children's understanding of regrouping.

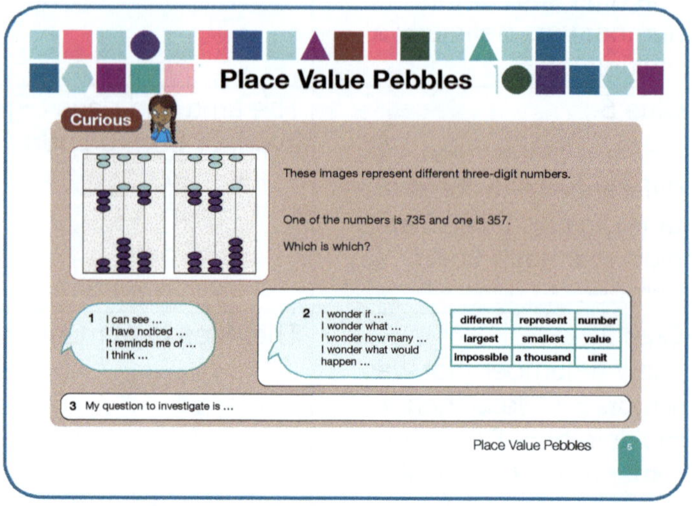

Children are shown images representing two numbers but are not told which image represents each number. They need to think about what is the same and what is different in order to work out the value of the pebbles in each section of the abacus.

Once children have discovered how the abacus works (a challenge in itself), there are several lines of enquiry they could follow. For example, they could work out the largest and smallest numbers that can be made; or explore whether there are different ways of representing the same number. They could even consider changing the values of the pebbles to create their own counting system.

This task promotes greater-depth learning in a range of ways, supporting children to identify and articulate underlying mathematical structures. For example, in the **Curious** stage, greater depth is promoted by asking children to make rich mathematical contributions to discussions and suggest a variety of appropriate mathematical questions.

This **Teacher Guide** is supported by photocopiable **Pupil Resources** and **Teacher's Presentation** PowerPoint slides for whole-class or group display.

National Curriculum

Place Value Pebbles covers these Year 4 National Curriculum objectives.

Number and place value

- Identify, represent and estimate numbers using different representations.
- Solve number and practical problems tht involve all of the above* and with increasingly large positive numbers.

*See National Curriculum statutory requirements for Year 4 Number and place value for the full list of statements to which this refers.

The 6C cycle

The grid on the right shows the relative weighting given to each of the 6 Cs in **Place Value Pebbles**.

Curious	★★★
Connect	★★
Collaborate	★★
Communicate	★★★
Create	★★★
Consider	★★

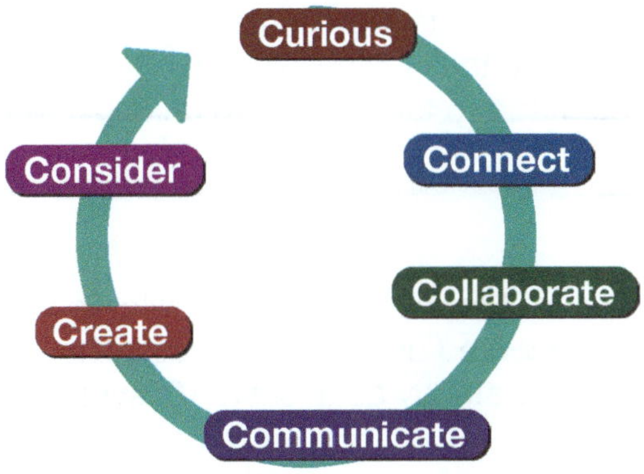

Problem-solving heuristics

The grid shows which problem-solving strategies children are likely to use when investigating the Connect tasks in **Place Value Pebbles**. These strategies are further exemplified in the Connect stage below (pages 39–40).

Draw a diagram	
Make a table / spot a pattern	✓
Make a model / act it out	✓
Try then improve	
Create a systematic list	✓
Solve a simpler but related problem	✓
Work backwards	

Key mathematical ideas

The abacus has three columns (for hundreds, tens and ones) and is split into an upper and lower section.

In the lower section, each bead or 'pebble' represents a unit of 100, 10 or 1, reading from left to right. To represent a number, pebbles in the bottom section are moved up to the horizontal separating line.

In the upper section, each pebble represents a unit of 500, 50 or 5. To represent a number, pebbles in the top section are moved down to the separating line. These pebbles are used to regroup the pebbles in the bottom section. For example, once five 10-pebbles in the bottom section are moved up, these can be replaced with a 50-pebble from the top section, and the five 10-pebbles can be moved back down. Moving a 10-pebble back up would then represent 60.

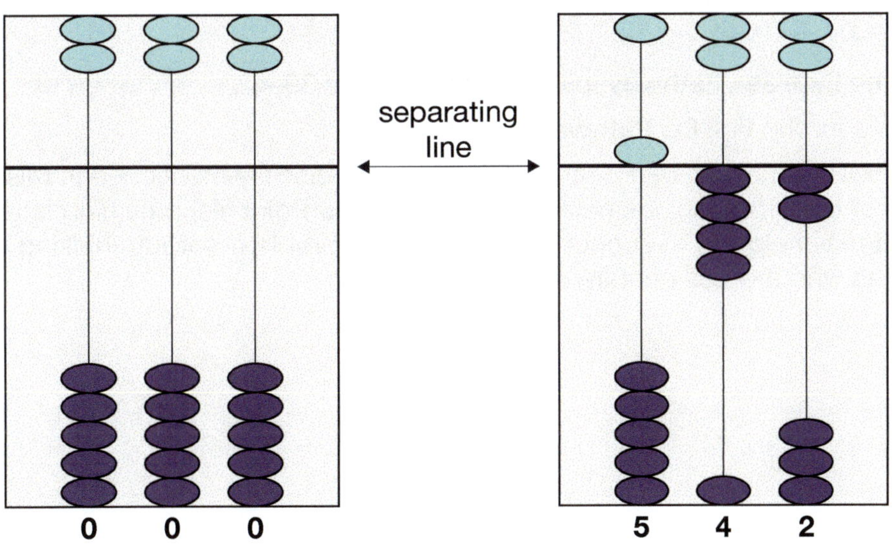

To make 542, two 1-pebbles and four 10-pebbles have been moved up, and then five 100-pebbles have been regrouped into one 500-pebble.

Exploring this representation encourages children to consider alternative representations of numbers and use the similarities with our base-10 number system to understand when to regroup. A link could be made to tallying where units of 5 are represented as:

One of the lines of enquiry children might choose to follow is: what is the highest value that can be represented using the abacus? The answer to this will depend on whether children decide to use the abacus strictly as a 'base-10' system, or if they choose to consider it as a 'base-15' system.

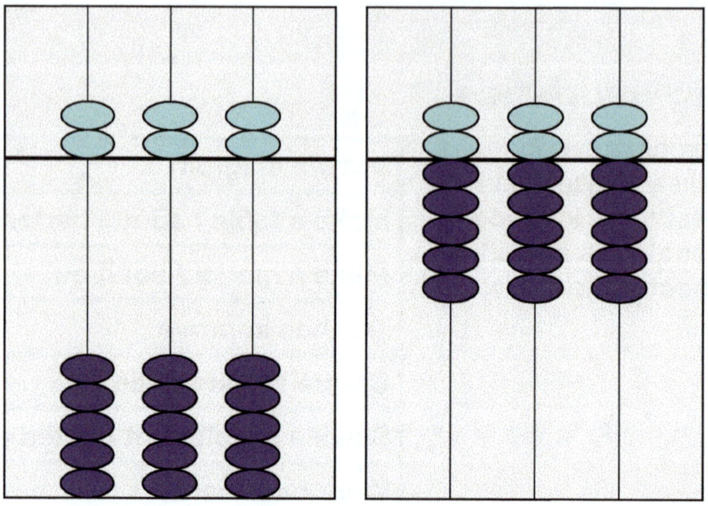

The numbers 1110 and 1665 represented using the abacus

In the image representing the number 1110, each set of five pebbles in the bottom section has been regrouped twice, with the pairs of pebbles in the top section being used to represent 10 ones, 10 tens, 10 hundreds. This is the highest value that can be represented if a 'base-10' rule is applied.

In the image representing 1665, an additional 5 ones, 5 tens and 5 hundreds are used (these cannot be regrouped as we do not have a third 500 / 50 / 5 pebble), making a total of 15 ones, 15 tens and 15 hundreds. This is the highest value that can be represented if a 'base-15' rule is applied.

Answers are given in situ throughout the tasks, and are also summarised on page 43.

Teaching guidance

If you are using the **Express Pathway**, please turn to pages 39–42.

Teaching guidance for the **Full 6C Pathway** is given below.

The **Teacher's Presentation** PowerPoint file has slides specifically mentioned in this teaching guidance. Some of the slides include notes (within the PowerPoint file) providing 'stumble support' suggestions, that may help you to support children in their problem-solving thinking and enable access for children who are less confident.

Awe and wonder

Reveal slides 1–2 one at a time, allowing time for discussion and observations as prompted below.

Slide 1

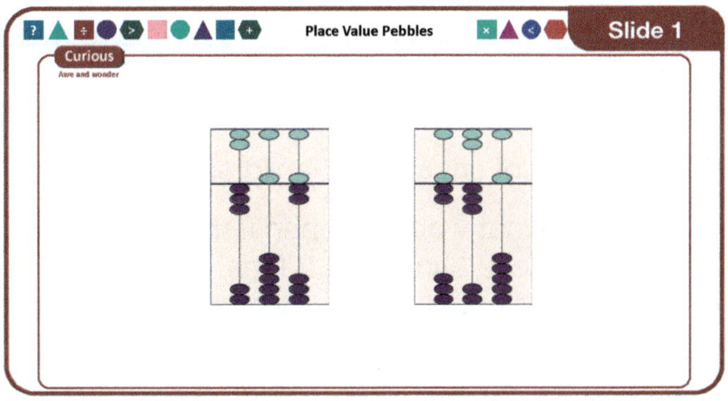

- I can see lots of pebbles / beads.
- I have noticed that they are in 3 sets / there are 3 columns.
- I have noticed that there is a horizontal line between the top section and the bottom section.
- I can see that there are 7 pebbles in each column, with 2 in the top section and 5 in the bottom.
- It reminds me of a wooden abacus.

Slide 2

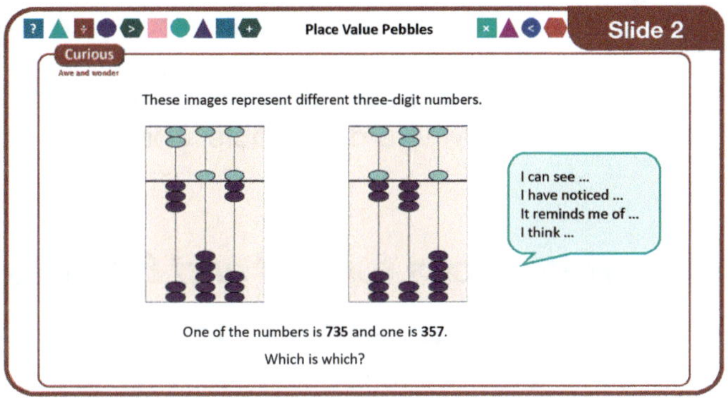

- *What do you notice?*
- *What questions do you have?*
- *What's the difference between the pebbles in the top section and the pebbles in the bottom section?*
- I think the first column is hundreds, the second column is tens and the third column is ones.
- I have noticed that some pebbles in the top section have been moved down to the line, and some pebbles from the bottom section have been moved up to the line.
- I see a group of 3 pebbles in the first column of the first image, so maybe that one is 357. But there is also a group of 3 pebbles in the first column of the second image.
- I see a group of 5 pebbles in the second column of the first image. In the second image, the second column doesn't have a group of 5. So I think the first image must be 357.

Develop and deepen

Children will get the most out of this Develop and deepen task once they have grasped how the abacus works. If it has not yet 'clicked' for children, move ahead to the first task in the **Connect** stage (1. Solve a simpler but related problem), and then return to Develop and deepen.

Slide 3

Display slide 3.

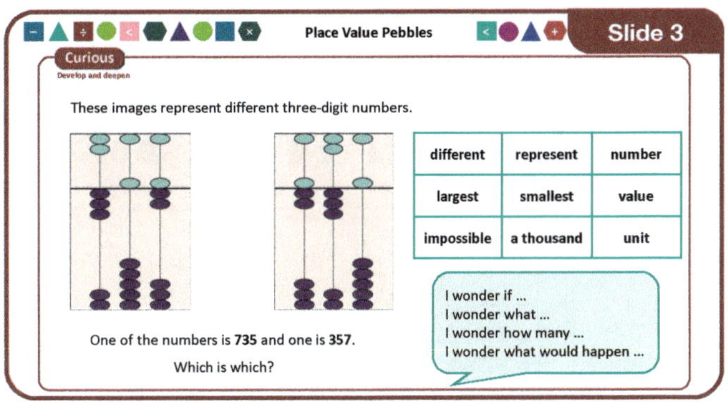

- Encourage children to use the words in the word bank to develop and deepen their thoughts and questions.

different	represent	number
largest	smallest	value
impossible	a thousand	unit

Possible lines of enquiry linked to the word bank:

- I wonder what different numbers I can represent.
- I wonder what's the highest number that I can represent.
- I wonder if any numbers are impossible to make.
- I wonder if I can make a thousand.
- I wonder how many different ways I can represent the same number.

Possible lines of enquiry that children may raise now, but should be held back until the **Create** task:

- I wonder what would happen if the pebbles in the bottom section represented a different unit, for example, 2 hundreds, 2 tens, 2 ones.
- I wonder if I could create my own counting system.

Consider whether / how to collate and display all the suggested questions, and whether / how to steer children towards selecting an appropriately challenging question.

Listing all of the possible lines of enquiry at this point would support children who are unsure of a route to take to choose one. It would also make pairing children up for investigation easier.

⚠ Watch out for children who are not confident in using the term 'unit' to describe each pebble (for example, where one pebble represents 50).

Pupil Resources for Full and Express Pathways

Distribute the **Pupil Resources** for **Place Value Pebbles** to the children to help support them in their exploration of the different tasks. Use either of the **Curious Pupil Resources** depending on whether you are following the **Full Pathway** or the **Express Pathway**. The **Pupil Resources** for the remaining 5Cs are common to both pathways.

To provide additional flexibility in the classroom, each of the **Pupil Resources** is also provided in the **Teacher's Presentation** PowerPoint file (see slides 4–7), so you can display these at any time if you wish.

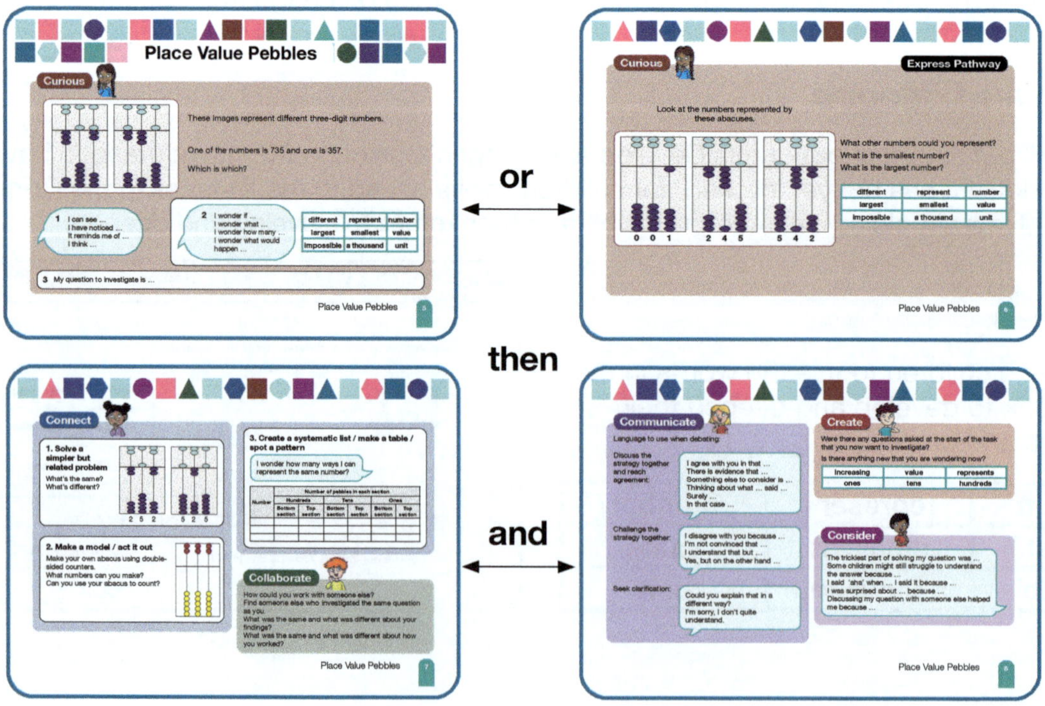

Express Pathway

If you are following the **Express Pathway**, spend a few minutes discussing the **Curious Pupil Resource** with the children. (This is slide 5, so you could display it if you wish.) Use some of the prompts on the previous page to help.

Three tasks involving different strategies are suggested for children to make connections.

1. Solve a simpler but related problem

Slide 8

Display slide 8, which shows two numbers that are easier to compare (252 and 525). For children who are yet to work out the value of the pebbles, this could be used before returning to the Develop and deepen task.

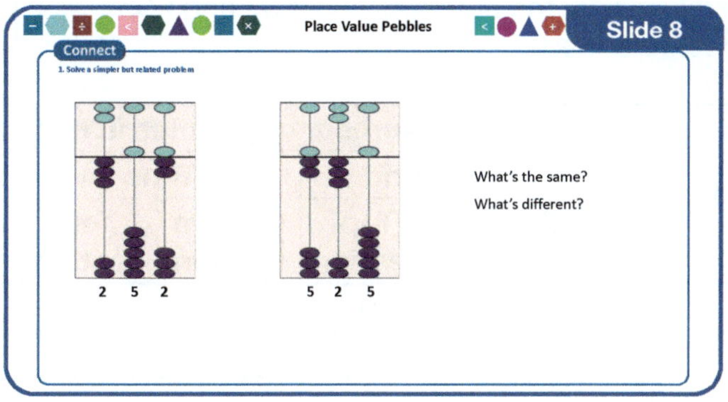

- *What's the same? What's different?*
- All the '2' digits are represented by 2 pebbles being moved up in the bottom section. All the '5' digits are represented by 1 pebble being moved down in the top section. So that must mean that the pebbles in the bottom section are worth 1, and the pebbles in the top section are worth 5. When a pebble is moved down or up to touch the separating line, that means it is 'in use'.

2. Make a model / act it out

Additional resources required: double-sided counters or counters in two different colours

Slide 9

Print slide 9 and distribute to children.

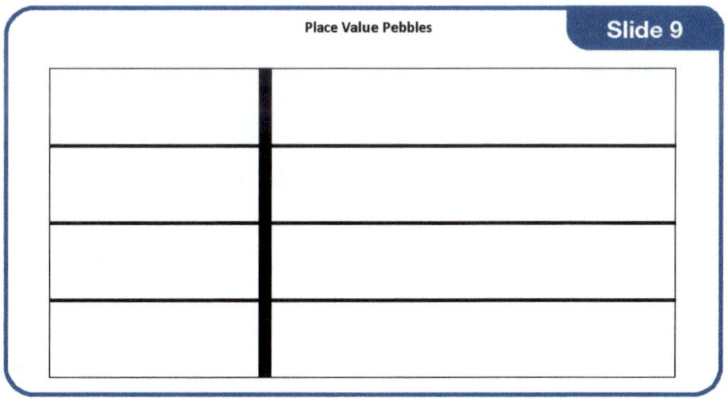

Slide 10

Display slide 10.

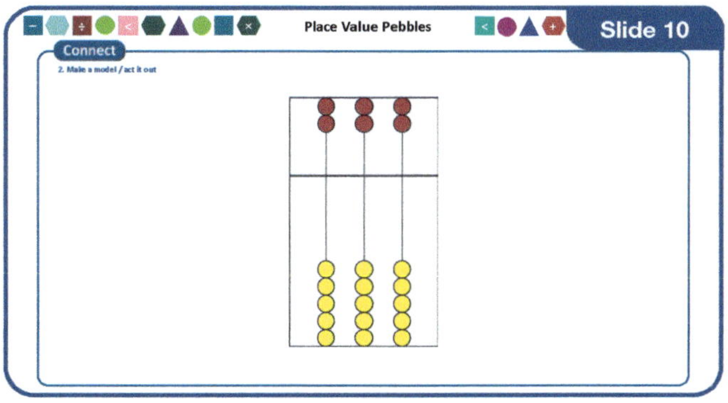

Children could make their own abacus, by lining up counters on the three vertical lines on the print-out of slide 9. Using different coloured counters in the top and bottom sections will help children to see the difference in unit values.

Encourage children to 'play' with their model, to represent a variety of numbers, and to count.

They could then use their abacus to work on their own lines of enquiry.

Connect

3. *Create a systematic list / make a table / spot a pattern*

Slide 11

Display slide 11.

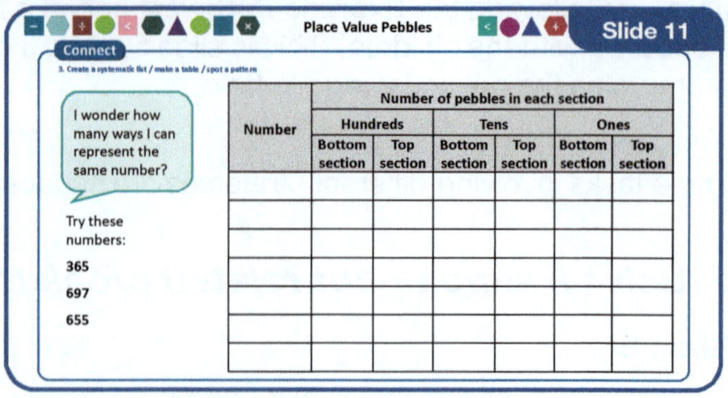

- I wonder how many ways I can represent the same number.

Children could use their own abacus (double-sided counters on a print-out of slide 9) to explore how the numbers 365, 697 and 655 could be made.

Number	Number of pebbles in each section					
	Hundreds		Tens		Ones	
	Bottom section	Top section	Bottom section	Top section	Bottom section	Top section
365	3	0	1	1	5	0
365	3	0	5	1	0	1
697	1	1	4	1	2	1
655	1	1	5	0	5	0
655	1	1	5	0	0	1
655	1	1	0	1	0	1
655	1	1	0	1	5	0

- *Which numbers can be made in more than one way?* (Numbers that contain a '5' digit.)

Collaborate

Refer to the **Collaborate** task of the **Pupil Resource**.

Pairs of children who worked on the same line of enquiry could compare their findings.

- *What question did you choose to investigate?*
- *What did you each find out? Were your answers the same?*
- *What strategy did you use? What did your partner do differently?*

You could choose to do the **Collaborate** task later in the task instead. If children create their own counting systems during the **Create** task, then 'Beat the timer' could be used. Children rotate after a set time to allow a variety of pairings or groupings. A timer could be used to monitor the timings and keep the session at a good pace.

Refer to the **Communicate** task of the **Pupil Resource**.

Slide 12

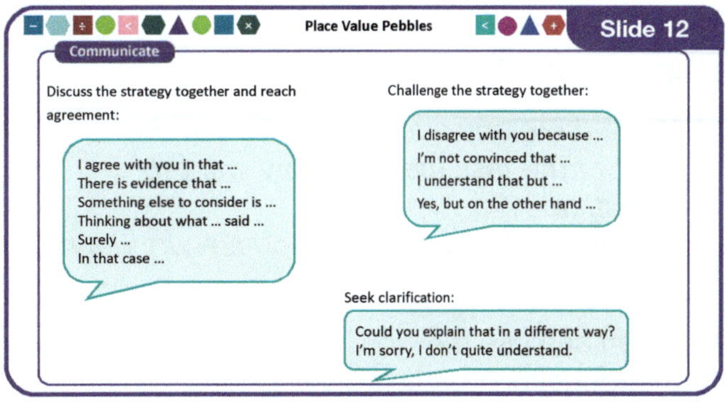

Display slide 12.

This task provides an opportunity for children to communicate their findings in a number of different ways. Children may need to be supported in choosing the most suitable way to communicate what they have found out.

Children could build their own abacus using materials such as a shoe box, string and beads, and demonstrate how it works. This could be presented verbally to a class, or children could make a short video to share on the school website.

Children could debate / write a persuasive argument about the highest number that could be created, or whether any numbers are impossible (the answer to these questions will depend on whether they decide to apply a base-10 or base-15 rule).

The following sentence stems might be useful to help children to frame their argument when debating.

Language for discussing the strategy together and reaching agreement:

- I agree with you in that …
- There is evidence that …
- Something else to consider is …
- Thinking about what … said …
- Surely …
- In that case …

Language for challenging the strategy together:

- I disagree with you because …
- I'm not convinced that …
- I understand that but …
- Yes, but on the other hand …

Language for seeking clarification:

- Could you explain that in a different way?
- I'm sorry, I don't quite understand.

Create

Refer to the **Create** task of the **Pupil Resource**.

Slide 13

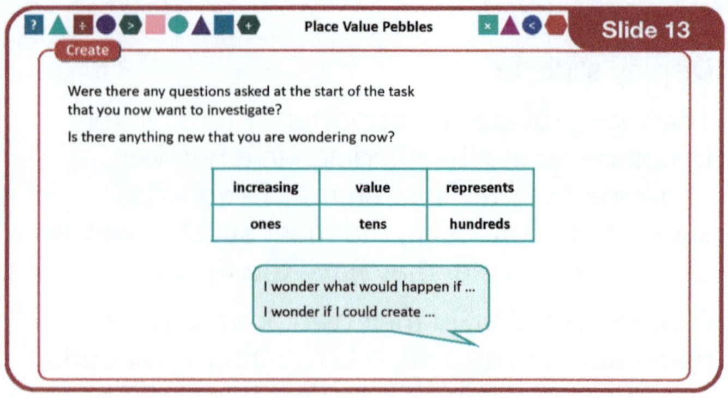

Display slide 13.

- *Were there any questions asked at the start of the task that you now want to investigate?*
- *Is there anything new that you are wondering now?*

Children can use the word bank to prompt further questions of their own.

increasing	**value**	**represents**
ones	**tens**	**hundreds**

- I wonder what would happen if the pebbles in the bottom section represented a different unit, for example, 2 hundreds, 2 tens, 2 ones.
- I wonder if I could create my own counting system.

Consider

Refer to the **Consider** task of the **Pupil Resource**.

Slide 14

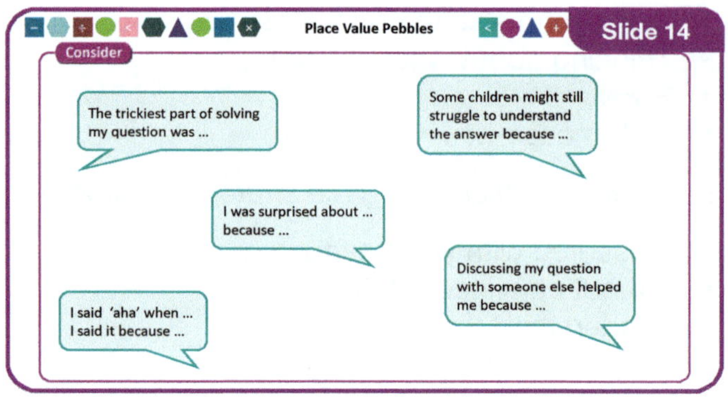

Display slide 14.

Encourage children to think about the strategies that they have used. The prompts below may help get children started.

Cognitive learning domain

- The trickiest part of solving my question was …
- I learned …
- Some children might still struggle to understand the answer because …

Affective domain

- The fun bit of my learning was … It was fun because …
- I said 'aha' when … I said it because …
- I was surprised about … because …

Metacognitive domain

- I got stuck when … I unstuck myself by …
- I used … to help me. I was helpful because …
- Discussing my question with someone else helped me because …

Answers

The lowest number that can be made is 0.

The highest number that can be made could be considered to be either:

- 1110 (when applying a base-10 rule to the abacus); or
- 1665 (when applying a base-15 rule).

Any number between 0 and 1110 (or 1665) can be represented.

When applying a base-10 rule, numbers with no '5' digit can only be represented in one way. Numbers with a '5' digit can be represented in more than one way, as a 5 can be made either using one pebble from the top section or five pebbles from the bottom section.

Changing the values of the pebbles

If the values of the pebbles are increased, the maximum total value that can be represented increases as follows, depending whether a base-10 or base-15 rule is applied. When the values of the pebbles change, some numbers are impossible to represent. For example, only even numbers can be represented if the pebbles in the bottom section are worth 2, 20 and 200.

Base-10 rule

As the value of the pebbles in the bottom section increases by 1, the maximum total value increases by 1110:

Values of pebbles in bottom section	Values of pebbles in top section	Maximum total value
1, 10, 100	5, 50, 500	1110
2, 20, 200	10, 100, 1000	2220
3, 30, 300	15, 150, 1500	3330
4, 40, 400	20, 200, 2000	6660

Children should be able to predict that, for example, if the values of the pebbles in the bottom section were 6, 60 and 600, the maximum total value would be 6660 ($6 \times 10 + 60 \times 10 + 600 \times 10$).

Base-15 rule

As the value of the pebbles in the bottom section increases by 1, the maximum total value increases by 1665:

Values of pebbles in bottom section	Values of pebbles in top section	Maximum total value
1, 10, 100	5, 50, 500	1665
2, 20, 200	10, 100, 1000	3330
3, 30, 300	15, 150, 1500	4995
4, 40, 400	20, 200, 2000	6660

Simple Subtraction

Year 3

Introduction

Simple Subtraction allows children to apply their mental subtraction knowledge and to seek out patterns within the number system.

During the **Curious** stage, children will discover that the difference between two 2-digit numbers (when the digits of the first number are reversed to make the second number) is always a multiple of 9. The task exemplifies how, through using a range of heuristics and working systematically, they can further explore the pattern and apply it to find pairs of numbers that fulfil the requirements.

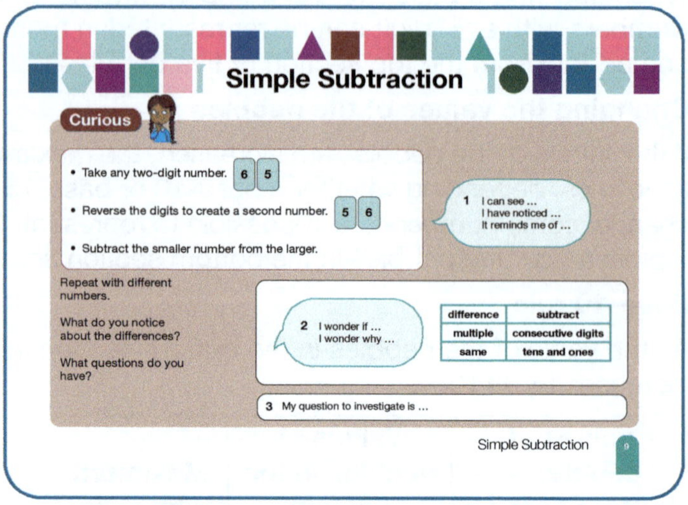

Children then compare and contrast the advantages and disadvantages of the different strategies, developing their ability to communicate through questioning and listening to each other. This task is rich in further creative opportunities to explore other operational patterns, applying their knowledge of an expanding range of strategies.

This task promotes greater-depth learning in a range of ways, supporting children to identify and articulate underlying mathematical structures. For example, in the **Connect** stage greater depth is promoted by asking children to explore a range of different mathematical resources and models, choose appropriate resources for the task, and explain their choice.

This **Teacher Guide** is supported by photocopiable **Pupil Resources** and **Teacher's Presentation** PowerPoint slides for whole-class or group display.

National Curriculum

Simple Subtraction covers these Year 3 National Curriculum objectives.

Addition and subtraction

- Add and subtract numbers mentally, including:
  - a three-digit number and ones;
  - a three-digit number and tens;
  - a three-digit number and hundreds.
- Add and subtract numbers with up to three digits, using formal written methods of columnar addition and subtraction.

The 6C cycle

The grid below shows the relative weighting given to each of the 6 Cs in **Simple Subtraction**.

Curious	★
Connect	★★★
Collaborate	★★
Communicate	★★★
Create	★★
Consider	★★

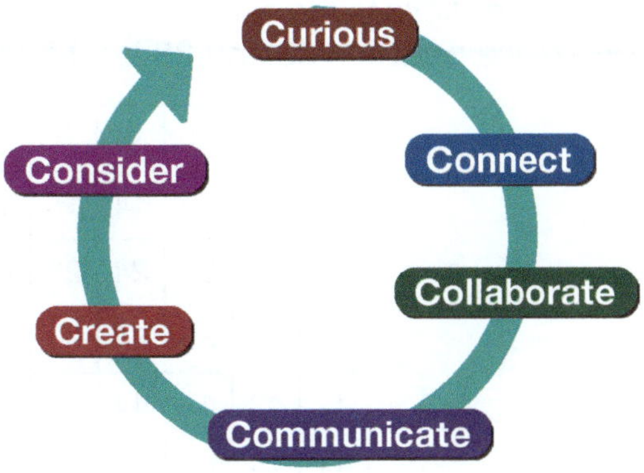

Problem-solving heuristics

The grid shows which problem-solving strategies children are likely to use when investigating the Connect tasks in **Simple Subtraction**. These strategies are further exemplified in the Connect stage below (pages 49–50).

Draw a diagram	
Make a table / spot a pattern	✓
Make a model / act it out	✓
Try then improve	
Create a systematic list	
Solve a simpler but related problem	
Work backwards	

Key mathematical ideas

Simple Subtraction provides the opportunity to explore the power and beauty of patterns within our number system. Children will use their mental subtraction skills to discover a pattern: the difference between two 2-digit numbers, where the digits are reversed in the second number, is always a multiple of 9.

Numbers with consecutive digits (for example, 12) produce a difference of 9 when they are reversed and the smaller is subtracted from the larger number (for example, $21 - 12 = 9$). This is because the tens digit in the smaller number is 1 less (subtract 10) but the ones digit is 1 more (add 1). So the difference between the numbers is 9. This can be seen by looking at these numbers on a 1–100 number square:

11	12
21	22

33	34
43	44

77	78
87	88

For numbers without consecutive digits (for example, 38 and 83), the difference between these numbers is always a multiple of 9. On the 1–100 number square below, there are 5 groups of 9 that separate 38 and 83. The difference between 38 and 83 is 45 which is the fifth multiple of 9. The difference between the digits is the same as the number of groups of 9 between the two numbers: $8 - 3 = 5$; the difference between 83 and 38 is 9 multiplied by 5.

1	2	3	4	5	6	7	8	9	10
11	12	13	14	15	16	17	18	19	20
21	22	23	24	25	26	27	28	29	30
31	32	33	34	35	36	37	38	39	40
41	42	43	44	45	46	47	48	49	50
51	52	53	54	55	56	57	58	59	60
61	62	63	64	65	66	67	68	69	70
71	72	73	74	75	76	77	78	79	80
81	82	83	84	85	86	87	88	89	90
91	92	93	94	95	96	97	98	99	100

$83 - 38 = 45$

Subtracting 45 is the same as subtracting 9 five times.

Some children may investigate numbers with the same digits, for example, $44 - 44$, which will result in a difference of 0. They may think this 'breaks' the rule, but 0 is still a multiple of 9 as $0 \times 9 = 0$.

This could be summarised algebraically, where A and B are the digits in the original number:

$$BA - AB = 9(B - A)$$

for example, $83 - 38 = 9 \times (8 - 3) = 9 \times 5 = 45$

Answers to the tasks are given on page 53.

Teaching guidance

If you are using the **Express Pathway**, please turn to pages 48–52.

Teaching guidance for the **Full 6C Pathway** is given below.

The **Teacher's Presentation** PowerPoint file has slides specifically mentioned in this teaching guidance. Some of the slides include notes (within the PowerPoint file) providing 'stumble support' suggestions, that may help you to support children in their problem-solving thinking and enable access for children who are less confident.

Awe and wonder

Slide 1

Display slide 1.

Give children time to explore a variety of calculations. Ask them to write each calculation on a sticky note or small piece of paper. This will later allow the calculations to be grouped (for example, all those with a difference of 9).

Display the calculations (not grouped at this stage) and give children time to look at them all.

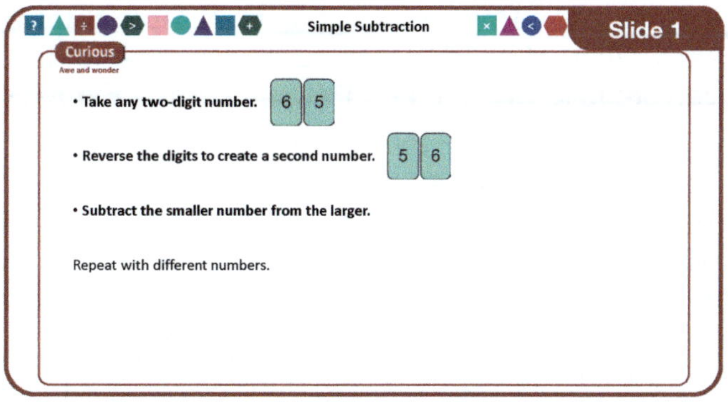

Slide 2

Display slide 2.

Give children time to discuss what they have noticed so far in the calculations.

- <u>I have noticed</u> that some of the differences are even numbers and some are odd.
- <u>I can see</u> that quite a few have differences of 9 / 18 / 27.
- <u>I have noticed</u> that all the calculations involved regrouping across the tens and ones.
- <u>It reminds me of</u> our work on deciding whether a mental or written method would be best.

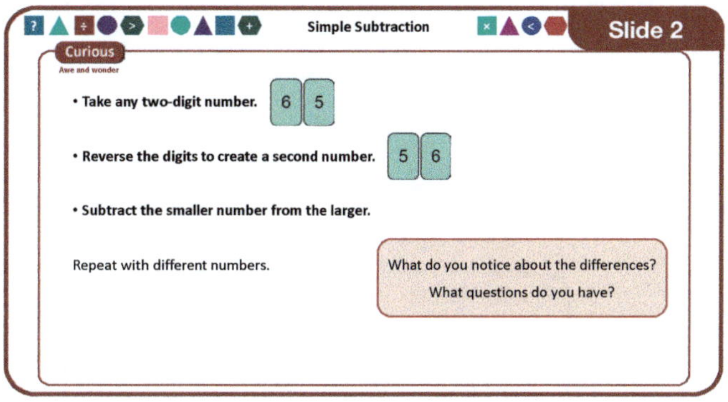

Develop and deepen

Slide 3

Display slide 3.

Encourage children to use the words in the word bank to develop and deepen their thoughts and questions.

difference	**subtract**	**multiple**
consecutive digits	**same**	**tens and ones**

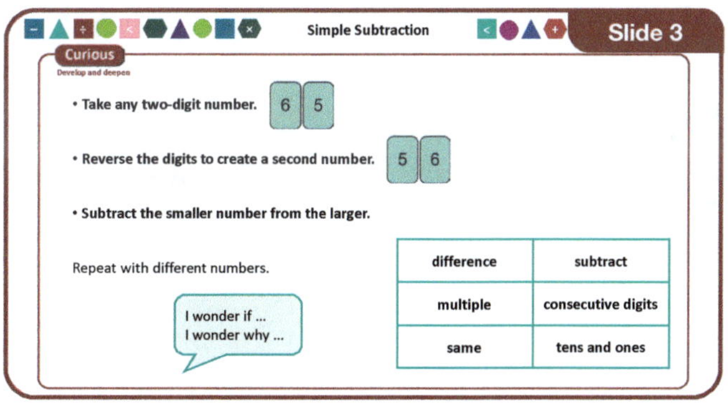

Possible lines of enquiry:

- <u>I wonder</u> why some of the differences are the same, for example, $65 - 56 = 9$ and $21 - 12 = 9$.

Possible lines of enquiry linked to the word bank:

- <u>I wonder</u> if all the differences have digits that add up to 9, for example, $1 + 8 = 9$ and $4 + 5 = 9$.
- <u>I wonder</u> if it's always true that the differences are all multiples of 9.
- <u>I wonder</u> if the difference will always be 9 if the two numbers have consecutive digits.
- <u>I wonder</u> if I could predict what the difference between a pair of numbers will be.

Possible lines of enquiry that children may raise now, but should be held back until the **Create** task:

- <u>I wonder</u> if the difference is a multiple of 9 when three-digit numbers are reversed and subtracted.

Consider whether / how to collate and display all the suggested questions, and whether / how to steer children towards selecting an appropriately challenging question.

⚠ Children may not be familiar with the term 'consecutive'.

Pupil Resources for Full and Express Pathways

Distribute the **Pupil Resources** for **Simple Subtraction** to the children to help support them in their exploration of the different tasks. Use either of the **Curious Pupil Resources** depending on whether you are following the **Full Pathway** or the **Express Pathway**. The **Pupil Resources** for the remaining 5Cs are common to both pathways.

To provide additional flexibility in the classroom, each of the **Pupil Resources** is also provided in the **Teacher's Presentation** PowerPoint file (see slides 4–7), so you can display these at any time if you wish.

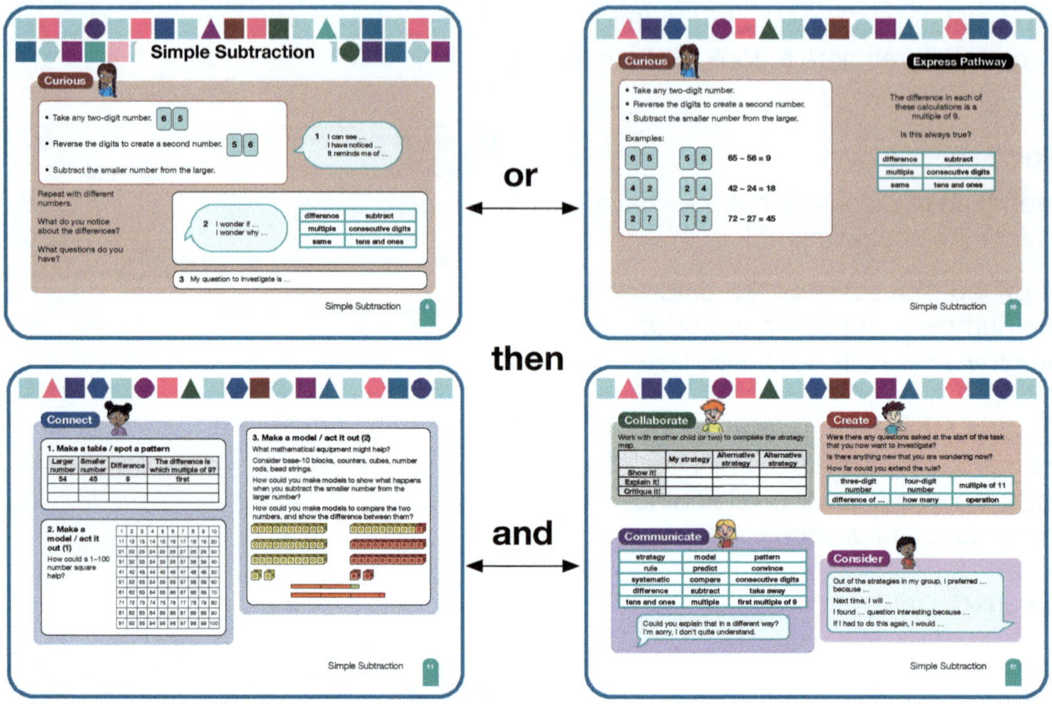

Express Pathway

If you are following the **Express Pathway**, spend a few minutes discussing the **Curious Pupil Resource** with the children. (This is slide 5, so you could display it if you wish.) Use some of the prompts on the previous page to help.

Three tasks involving different strategies are suggested for children to make connections.

1. Make a table / spot a pattern

Slide 8

Print slide 8 and distribute to children.

Ask children to investigate the relationship between the pairs of numbers and their differences by organising their workings into a table. Encourage them to work systematically.

This will help children to spot the pattern that the difference produced from subtracting the smaller number from the larger number is always a multiple of 9. Identifying this may lead them to notice further patterns and rules.

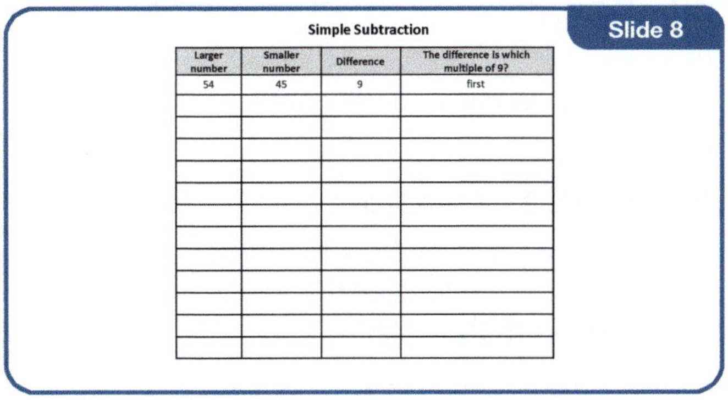

Larger number	Smaller number	Difference	The difference is which multiple of 9?
54	45	9	first
92	29	63	seventh
75	57	18	second
53	35	18	second

2. Make a model / act it out (1)

Slide 9

Print slide 9 and distribute to children. Ask them to colour the pairs of numbers in different colours on the 1–100 number square to help them to spot the patterns.

Children may notice that the number of rows that separate the numbers is the same as the number of columns. For example, 24 and 42 are two rows and two columns apart. Once this has been spotted, the next step is to notice that this means that multiples of 9 create the difference between each pair. For example:

$$24 + 10 - 1 + 10 - 1 = 42$$
$$24 + 9 + 9 = 42$$
$$24 + 18 = 42$$

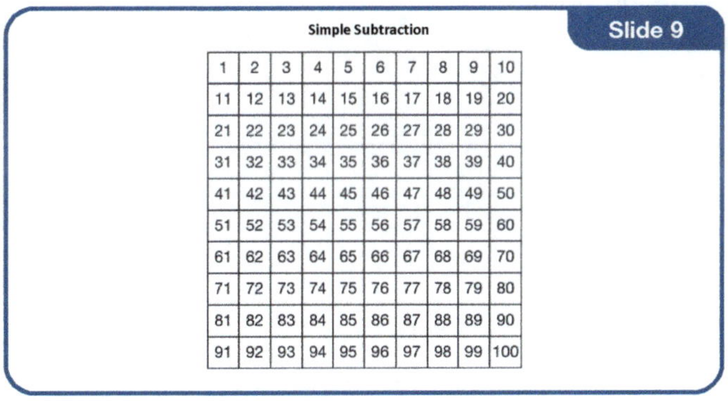

Slide 10

Display slide 10 if children are struggling with accurate subtraction.

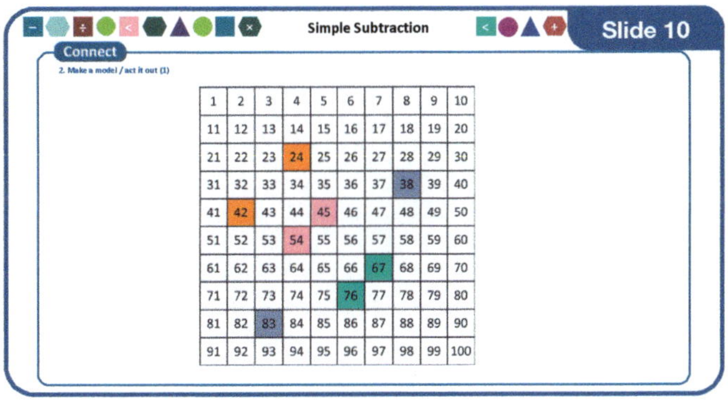

3. Make a model / act it out (2)

Additional resources required: base-10 blocks, counters, cubes, number rods, bead strings, Numicon Shapes

Using concrete resources will allow children to explore the structure of the numbers and expose why the number left over is a multiple of 9.

Ask children to think about what resources they might find helpful, and how they could use them.

Slide 11

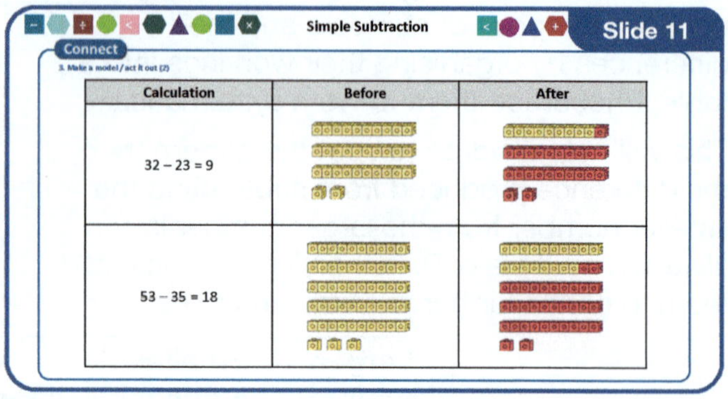

Display slide 11 if needed.

Some children may choose to model what happens when the smaller number is subtracted from the larger.

For example, children could use cubes in two different colours to make 'before' and 'after' models to reveal the structure of the calculation.

Some children may choose to make models that focus on comparing the two numbers and 'finding the difference'.

For example, they could build the two numbers by placing number rods end-to-end, with one model aligned below the other, then find a rod (or rods) to match the difference between them.

Slide 12

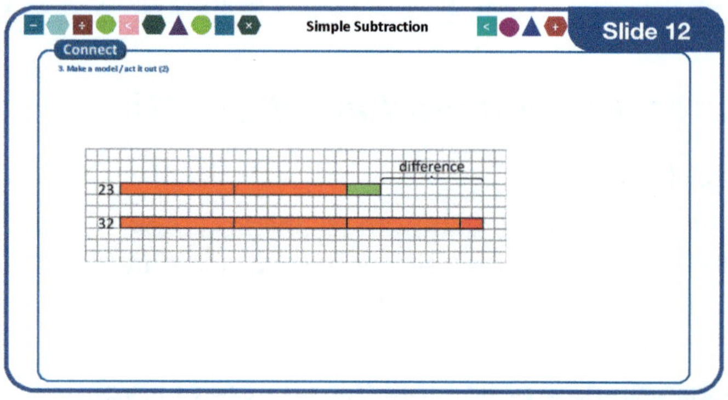

Display slide 12 if needed.

Ask children to build the numbers 23 and 32 using number rods, then find a rod that matches the difference between the two numbers (a 9-rod).

Slide 13

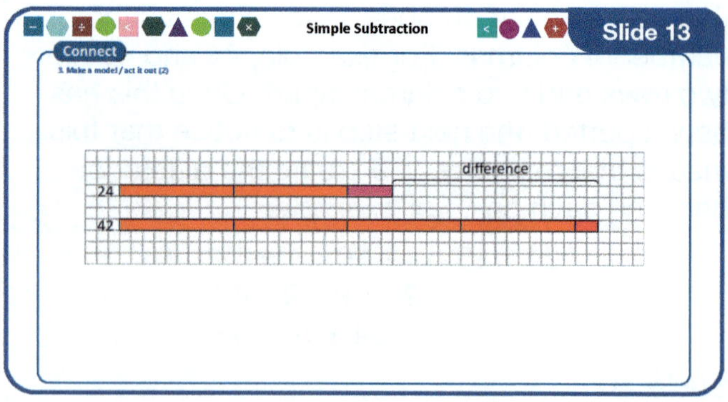

Display slide 13 if needed.

Ask children to build the numbers 24 and 42 using number rods, then find rods that match the difference between the two numbers (two 9-rods).

Refer to the **Collaborate** task of the **Pupil Resource**.

Slide 14

Print slide 14 and distribute to children.

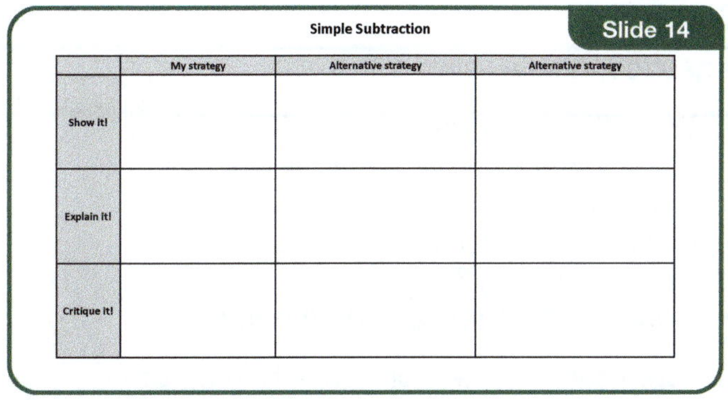

Ask each child to complete a strategy map, using the table as a template. (The strategy map could be adapted to include more or fewer strategies.)

This would work well in pairs or groups of three. For this task, organise children according to their preferred strategies from the **Connect** stage. In their pairs or groups, children each present their chosen strategy, then record details of the methods used in their group.

Children will need to collaborate closely to decide how best to articulate and explain their chosen strategy in order for other children to understand and then summarise it on their strategy map.

- *How could you explain your preferred strategy as clearly as possible?*
- *What questions could you ask your partner to help you understand their strategy?*
- *Which strategy do you now think is best? Why? Do you all agree?*

Refer to the **Communicate** task of the **Pupil Resource**.

This task provides an opportunity for children to present their preferred strategy to the class.

Once pairs/groups have completed their strategy maps, it is time for children to compare and contrast the different strategies by communicating their findings to the rest of the class. Children may choose to share their models, jottings and notes and to justify their chosen approach.

They can use the word bank for vocabulary support.

strategy	**model**	**pattern**
rule	**predict**	**convince**
systematic	**compare**	**consecutive digits**
difference	**subtract**	**take away**
tens and ones	**multiple**	**first multiple of 9**

As well as communicating their strategy, the second important part of this task's communication opportunities is to listen to, interpret and gain an understanding of the strategies presented by other groups. This may include having to seek clarification, by using sentence stems such as:

- Could you explain that in a different way?
- I'm sorry, I don't quite understand.

Create

Refer to the **Create** task of the **Pupil Resource**.

Slide 15

If needed, display slide 15.

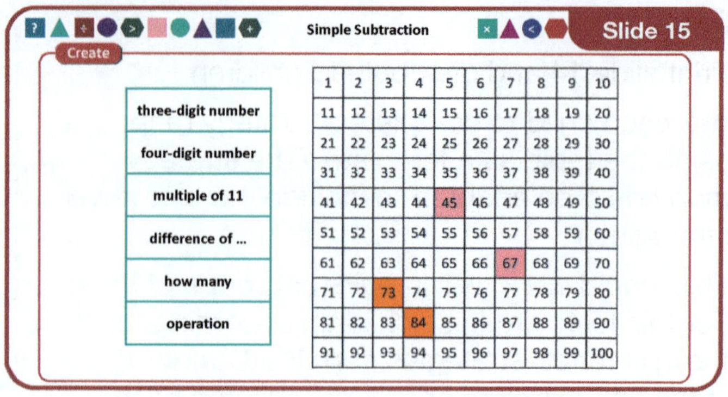

- *Were there any questions asked at the start of the task that you now want to investigate?*
- *Is there anything new that you are wondering now?*
- *How far could you extend the rule?*

Children can use the word bank to prompt further questions of their own.

three-digit number	**four-digit number**	**multiple of 11**
difference of …	**how many**	**operation**

- I wonder if the difference is a multiple of 9 when three-digit numbers are reversed and subtracted.
- I wonder if this is true for four-digit numbers.
- I wonder what happens for other operations, for example, 12×21, 32×23.
- I wonder how many pairs of numbers have a difference of 9 / 18 / … / 63.
- I wonder whether I can make the pattern go in the other direction. (Differences that are multiples of 11, rather than multiples of 9.)

Slide 15 shows a 1–100 number square with two pairs of numbers coloured: 73 and 84, which have a difference of 11; and 45 and 67, which have a difference of 22.

Consider

Refer to the **Consider** task of the **Pupil Resource**.

Encourage children to think about the strategies that they have used. The prompts below may help get children started.

Cognitive learning domain

- Out of the strategies in my group, I preferred … because…
- Next time, I will …
- I noticed that …

Affective domain

- I found … question interesting because …
- My aha moment was …
- I tried to influence the class through …

Metacognitive domain

- I found … helpful.
- I found … unhelpful.
- If I had to do this again, I would …

Answers

When taking a two-digit number, reversing the digits and finding the difference between the numbers, the answer is always a multiple of 9.

Calculations can be grouped according to the difference between the digits in the original numbers.

B – A = 1
BA – AB = 9

AB	BA – AB
01	10 – 01
12	21 – 12
23	32 – 23
34	43 – 34
45	54 – 45
56	65 – 56
67	76 – 67
78	87 – 78
89	98 – 89

Pattern:

When the difference between A and B is 1 (that is, the digits are consecutive), the difference between AB and BA is always 9 (the first multiple of 9).

B – A = 2
BA – AB = 18

AB	BA – AB
02	20 – 02
13	31 – 13
24	42 – 24
35	53 – 35
46	64 – 46
57	75 – 57
68	86 – 68
79	97 – 79

Pattern:

When the difference between A and B is 2, the difference between AB and BA is always 18 (the second multiple of 9).

B – A = 3
BA – AB = 27

AB	BA – AB
03	30 – 03
14	41 – 14
25	52 – 25
36	63 – 36
47	74 – 47
58	85 – 58
69	96 – 69

Pattern:

When the difference between A and B is 3, the difference between AB and BA is always 27 (the third multiple of 9).

Extending the pattern:

B – A = 4 BA – AB = 36 (fourth multiple of 9)
B – A = 5 BA – AB = 45 (fifth multiple of 9)
B – A = 6 BA – AB = 54 (sixth multiple of 9)
B – A = 7 BA – AB = 63 (seventh multiple of 9)

Generalising the pattern:

In order for a pair of ‘digit reversed’ two-digit numbers to have a difference of the *n*th multiple of 9, the difference between the digits in the numbers must be *n*.

X Marks the Spot

Year 4

Introduction

X Marks the Spot uses a simple cross design to dig deeply into number manipulation. It provides the opportunity for children to rehearse, and increase their fluency in, addition of sets of numbers less than 20, applying mental and written methods as appropriate. The task is an adaption of a magic square and involves making all the arms of the cross add up to the same value.

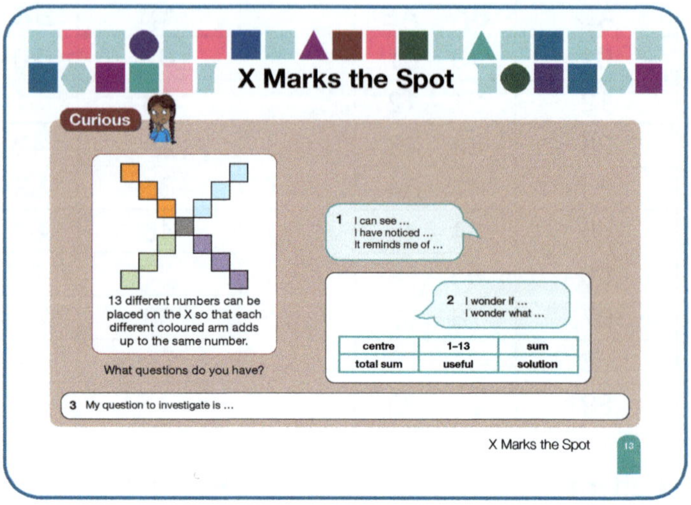

The focus of the task moves to exploring and comparing strategies through collaboration. Children explain their strategies, and develop questioning skills as they consider the strategies of others, before concluding which strategy is the most efficient for them. Children will recognise that different people find different strategies more efficient and no single strategy is 'best'.

This task promotes greater-depth learning in a range of ways, supporting children to identify and articulate underlying mathematical structures. For example, in the **Collaborate** stage, greater depth is promoted by asking children to offer suggestions, listen carefully to others and build upon their own ideas, and be able to accept it if their ideas are rejected by the group.

This **Teacher Guide** is supported by photocopiable **Pupil Resources** and **Teacher's Presentation** PowerPoint slides for whole-class or group display.

National Curriculum

X Marks the Spot covers these Year 4 National Curriculum objectives.

Addition and subtraction

- Add and subtract numbers with up to four digits using the formal written methods of columnar addition and subtraction where appropriate.
- Estimate and use inverse operations to check answers to a calculation.
- *Pupils continue to practise both mental methods and columnar addition and subtraction with increasingly large numbers to aid fluency.* [non-statutory guidance]

The 6C cycle

The grid below shows the relative weighting given to each of the 6 Cs in **X Marks the Spot**.

Curious	★★
Connect	★★★
Collaborate	★★★
Communicate	★★
Create	★★
Consider	★★

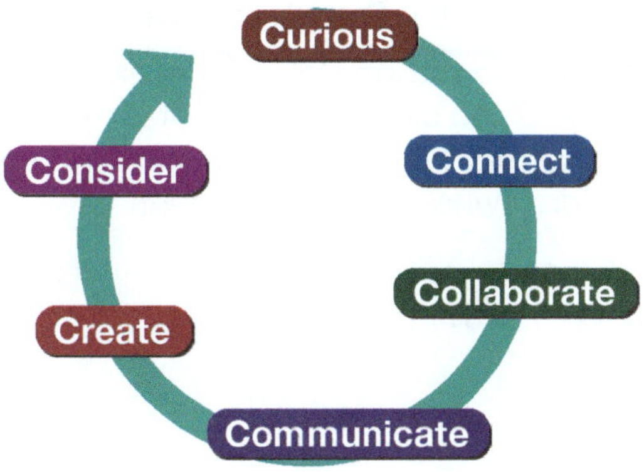

Problem-solving heuristics

The grid shows which problem-solving strategies children are likely to use when investigating the Connect tasks in **X Marks the Spot**. These strategies are further exemplified in the Connect stage below (pages 58–59).

Draw a diagram	
Make a table / spot a pattern	✓
Make a model / act it out	
Try then improve	✓
Create a systematic list	✓
Solve a simpler but related problem	✓
Work backwards	✓

Key mathematical ideas

X Marks the Spot takes simple mental calculation to a deeper level. It involves children seeking patterns and generalising relationships between numbers. Children are presented with a cross constructed from 13 squares, in which they have to place the numbers from 1 to 13 in such a way as to make each of the four arms sum to the same total.

Children should be encouraged to consider suitable starting points to solve the puzzle. Some children may opt for a trial and improvement strategy, which is detailed below in the Connect stage.

However, through the exploration of this problem, children may discover that if they can make the total for all four arms (excluding the centre square) a multiple of 4, this can then be divided by 4 to find the sum of each arm.

The total sum of the numbers 1–13 is 91.

$91 - 3 = 88$	$88 \div 4 = 22$	Each arm will sum to 22.
$91 - 7 = 84$	$84 \div 4 = 21$	Each arm will sum to 21.
$91 - 11 = 80$	$80 \div 4 = 20$	Each arm will sum to 20.

Children may choose to explore whether this rule would work for any 13 consecutive, or non-consecutive, numbers. They should find that the rule works for any 13 consecutive numbers, and the X can always be completed successfully.

For non-consecutive numbers, a solution is not always possible.

Answers to the tasks are given on pages 61–62.

Teaching guidance

If you are using the **Express Pathway**, please turn to pages 57–61.

Teaching guidance for the **Full 6C Pathway** is given below.

The **Teacher's Presentation** PowerPoint file has slides specifically mentioned in this teaching guidance. Some of the slides include notes (within the PowerPoint file) providing 'stumble support' suggestions, that may help you to support children in their problem-solving thinking and enable access for children who are less confident.

Awe and wonder

Slide 1

Display slide 1.

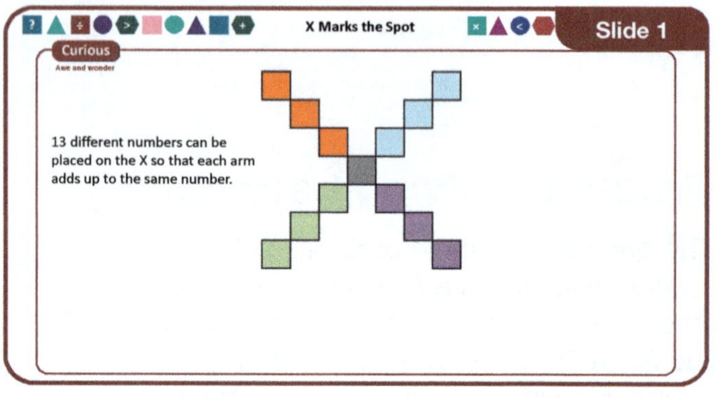

- I can see that the squares make an X shape.
- I have noticed that each arm of the X is a different colour.
- I can see that there is a square in the centre of the X and all of the arms are joined to it.
- I have noticed that it says 13 different numbers.
- I have noticed that there are 13 squares in total.
- I have noticed that each arm has 3 squares and that there are 4 arms.
- It reminds me of a magic square.

Develop and deepen

Slide 2

Display slide 2.

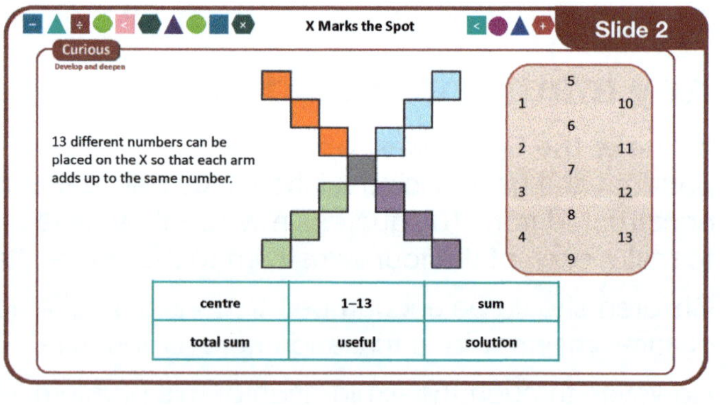

Encourage children to use the words in the word bank to develop and deepen their thoughts and questions.

centre	**1–13**	**sum**
total sum	**useful**	**solution**

Possible lines of enquiry:

- I wonder what all of the numbers add up to.

Possible lines of enquiry linked to the word bank:

- I wonder if we should use the numbers 1 to 13.
- I wonder if the centre number is important.
- I wonder if calculating the total sum will be useful.
- I wonder if there is more than one solution

Possible lines of enquiry that children may raise now, but should be held back until the **Create** task:

- I wonder if this would work with any 13 different numbers.
- I wonder what would happen if there were more squares on each arm.

Consider whether / how to collate and display all the suggested questions, and whether / how to steer children towards selecting an appropriately challenging question.

Pupil Resources for Full and Express Pathways

Distribute the **Pupil Resources** for **X Marks the Spot** to the children to help support them in their exploration of the different tasks. Use either of the **Curious Pupil Resources** depending on whether you are following the **Full Pathway** or the **Express Pathway**. The **Pupil Resources** for the remaining 5Cs are common to both pathways.

To provide additional flexibility in the classroom, each of the **Pupil Resources** is also provided in the **Teacher's Presentation** PowerPoint file (see slides 3–6), so you can display these at any time if you wish.

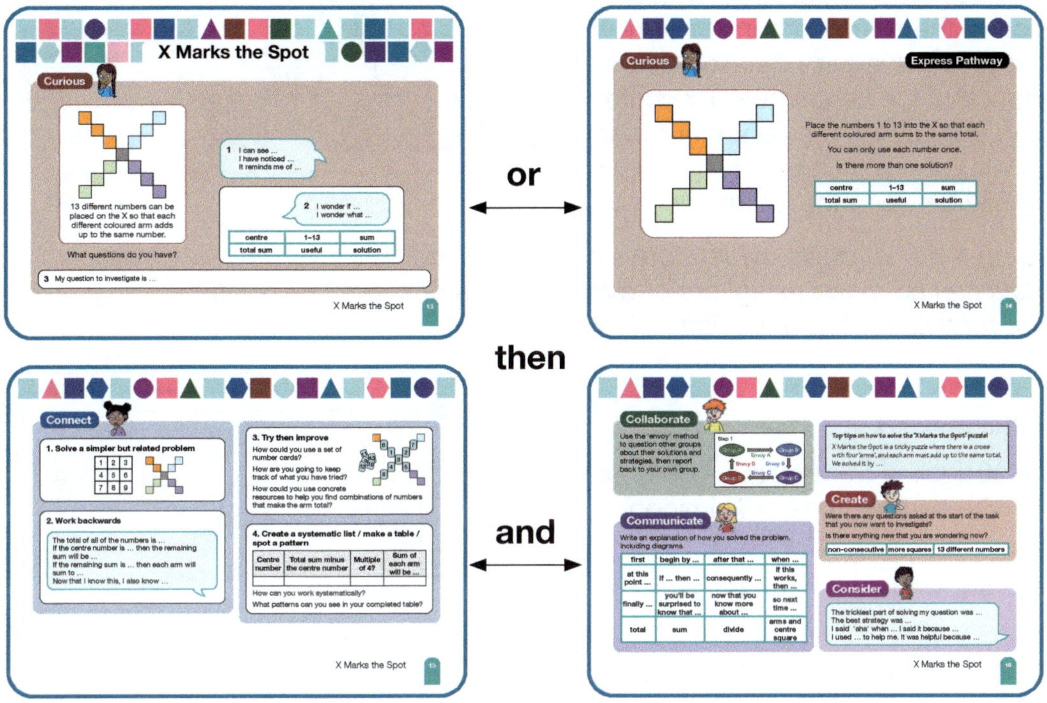

Express Pathway

If you are following the **Express Pathway**, spend a few minutes discussing the **Curious Pupil Resource** with the children. (This is slide 4, so you could display it if you wish.) Use some of the prompts on the previous page to help.

Four tasks involving different strategies are suggested for children to make connections.

1. Solve a simpler but related problem

Slide 7

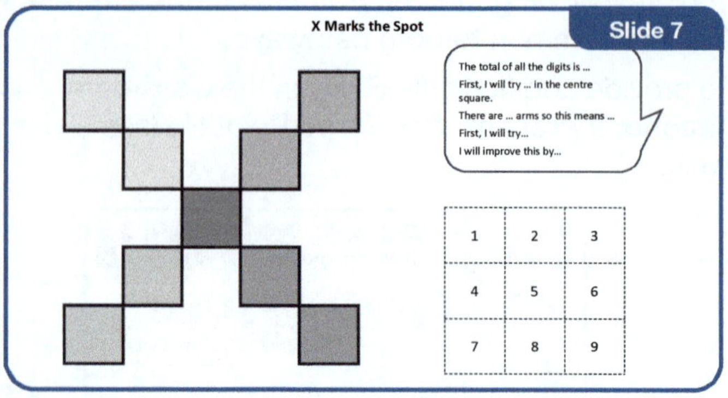

Print slide 7 and distribute to children. They could cut out the digit cards so they can be placed and rearranged on the X.

Ask children to use the numbers 1 to 9 on a smaller version of the X. Trial and improvement could be used as a strategy as detailed below, or the importance of the centre number could be explored. In this case, if the centre number is 9, 5 or 1, this will leave a multiple of 4 (36, 40 or 44) to then divide between the four arms.

- *When adding a pair of digits, which totals can you make?*
- *Which totals can be made using lots of different pairs of digits?* (9, 10, 11)
- *What is the total sum of all nine digits?* (45)

Slide 8

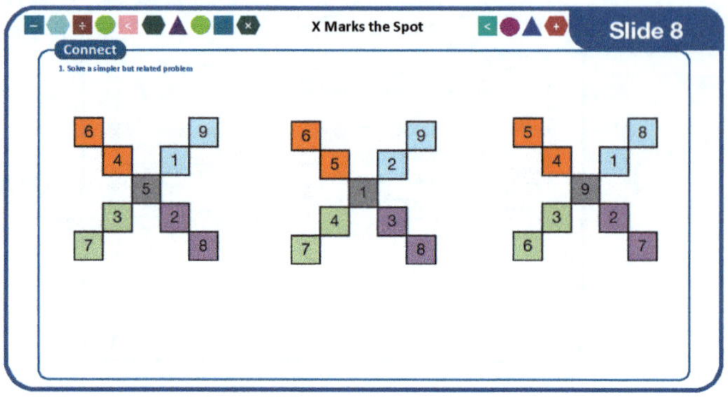

Display slide 8 if needed.

If children are struggling to calculate, you could provide them with number rods or Numicon Shapes.

You could also show them the three solutions on slide 8.

Children may misuse the term 'sum' to describe any calculation rather than the correct meaning of adding two or more numbers.

2. Work backwards

Children could start by totalling all of the numbers.

The following speaking frame may be useful:

- The total of all of the numbers is …
- If the centre number is … then the remaining sum will be …
- If the remaining sum is … then each arm will sum to …
- Now that I know this, I also know …

3. *Try then improve*

Additional resources required: number cards 1–13 (or sticky notes or another alternative), number rods or bead strings

Children may initially choose to try and then improve to find solutions. Number cards will make it easier to try different arrangements.

Encourage children to record their attempted arrangements to avoid replications.

If using 'try then improve' as a strategy along with another, encourage children to not disregard each arrangement that they make, but instead make one improvement each time.

When children have ascertained that the four arms will sum to a multiple of 4, and have identified the total sum of the four arms, number rods or bead strings could be used to explore different combinations of numbers to make the 'arm' total.

⚠ Watch out for children who have not yet realised that multiples of 4 will be useful.

- *What if there was no centre square, so only 12 'arm' squares? What would this tell you?*
- *Why might multiples of 4 be useful?*
- *What is the total of all 13 numbers?* (91)
- *Is this a multiple of 4?* (No.)

4. *Create a systematic list / make a table / spot a pattern*

Slide 9

Display slide 9 if needed.

This strategy would be useful once children have ascertained, or been prompted, that the total of all of the arms without the centre square should be a multiple of 4.

Completing this table will reveal that the centre number should be either 11, 7 or 3.

Trial and improvement could then be used to discover the positioning of the other digits.

- *Have you noticed any patterns in your table?*
- I notice that I didn't have to do most of the subtraction calculations. Each time the number I was taking away (subtrahend) got 1 smaller, the answer (difference) became 1 larger.

Children may not be familiar with the terms 'minuend' (number being reduced), and 'subtrahend' (number being subtracted). You may wish to introduce these terms to aid the clarity of children's explanations.

Centre number	Total sum minus the centre number	Multiple of 4?	Sum of each arm will be ...
13	91 – 13 = 78	No	
12	91 – 12 = 79	No	
11	91 – 11 = 80	Yes	20
10	91 – 10 = 81	No	
9	91 – 9 = 82	No	
8	91 – 8 = 83	No	
7	91 – 7 = 84	Yes	21
6	91 – 6 = 85	No	
5	91 – 5 = 86	No	
4	91 – 4 = 87	No	
3	91 – 3 = 88	Yes	22
2	91 – 2 = 89	No	
1	91 – 1 = 90	No	

Collaborate

Refer to the **Collaborate** task of the **Pupil Resource**.

Slide 10

Display slide 10.

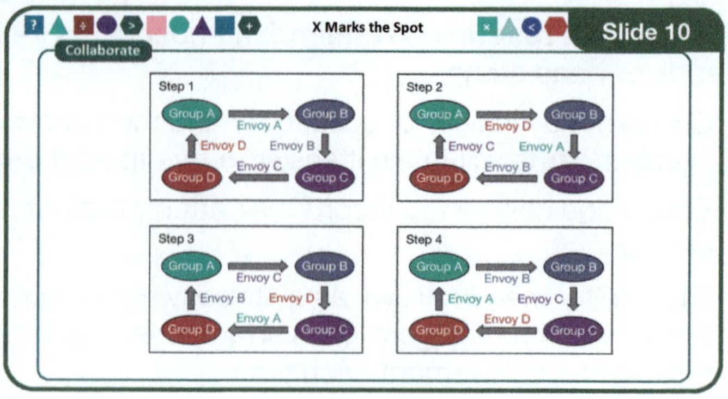

As there are a number of solutions to this puzzle, this task provides a good opportunity for children to use the 'envoy' method. One child from each group is nominated as the envoy. Following discussion and exploration time, the envoy then moves to each of the other groups in turn, to ask questions about how they tackled the problem and the solutions they found, before reporting back to their original group.

The envoys could ask the other groups questions such as:

- How did you work out how to make each arm add up to 20 / 21 / 22?
- Did you use any concrete resources to help you?
- Did you come up with a system for how to place the numbers?

Communicate

Refer to the **Communicate** task of the **Pupil Resource**.

Slide 11

Display slide 11.

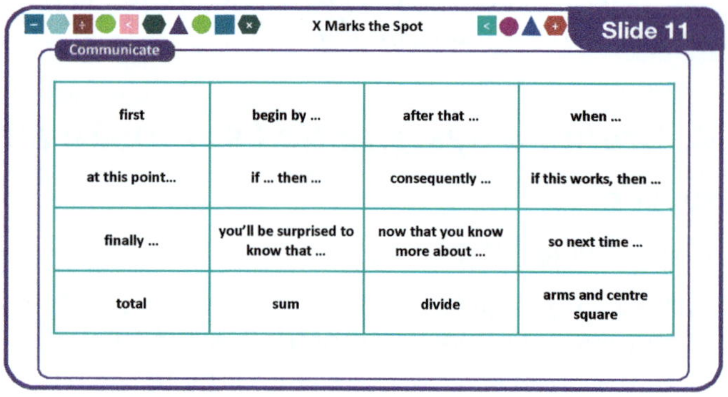

This task provides an opportunity for children to write an explanation text to show how they found one or more solutions to the puzzle.

Ask children to:

- address the reader directly;
- ensure that sentences or notes are sequenced logically;
- use diagrams such as flow charts to organise the stages of the process they followed.

Create

Refer to the **Create** task of the **Pupil Resource**.

Slide 12

Display slide 12.

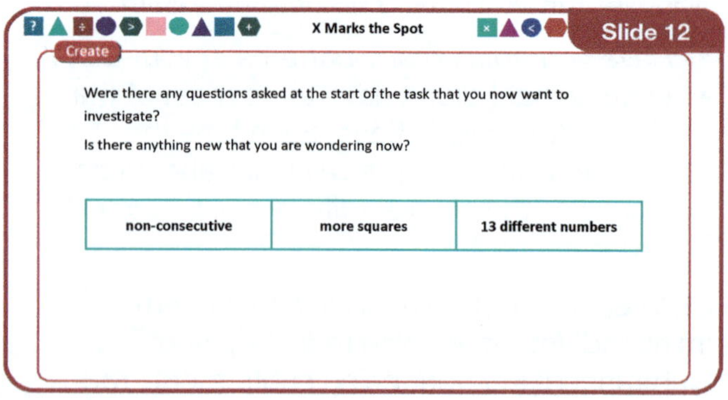

- *Were there any questions asked at the start of the task that you now want to investigate?*
- *Is there anything new that you are wondering now?*

Children can use the word bank to prompt further questions of their own.

non-consecutive	**more squares**	**13 different numbers**

- I wonder if this would work with 13 different numbers.
- I wonder if I could find a solution for 13 non-consecutive numbers.
- I wonder what would happen if there were more squares on each arm.

Refer to the Consider task of the **Pupil Resource**.

Slide 13

Display slide 13.

Encourage children to think about the strategies that they have used. The prompts below may help get children started.

Cognitive learning domain

- The trickiest part of solving my question was …
- I learned …
- The best strategy was …

Affective domain

- The fun bit of my learning was … It was fun because …
- I said 'aha' when … I said it because …
- I was surprised about … because …

Metacognitive domain

- I got stuck when … I unstuck myself by …
- I used … to help me. It was helpful because …
- Next time, I will …

Answers

If the numbers 1–13 are placed in the X, the centre number must be either 3, 7 or 11.

Centre number	Total for each arm		Example solution
	(excluding centre number)	(including centre number)	
3	$91 - 3 = 88$ $88 \div 4 = \mathbf{22}$	$22 + 3 = \mathbf{25}$	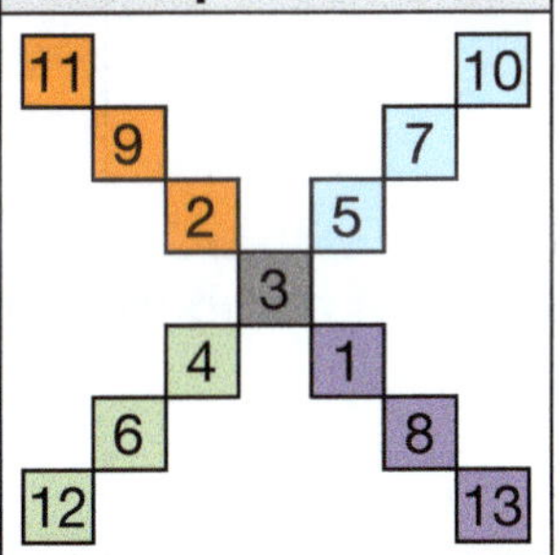
7	$91 - 7 = 84$ $84 \div 4 = \mathbf{21}$	$21 + 7 = \mathbf{28}$	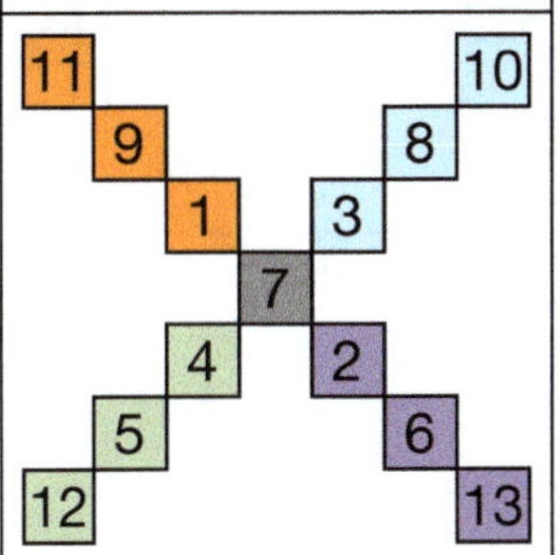
11	$91 - 11 = 80$ $80 \div 4 = \mathbf{20}$	$20 + 11 = \mathbf{31}$	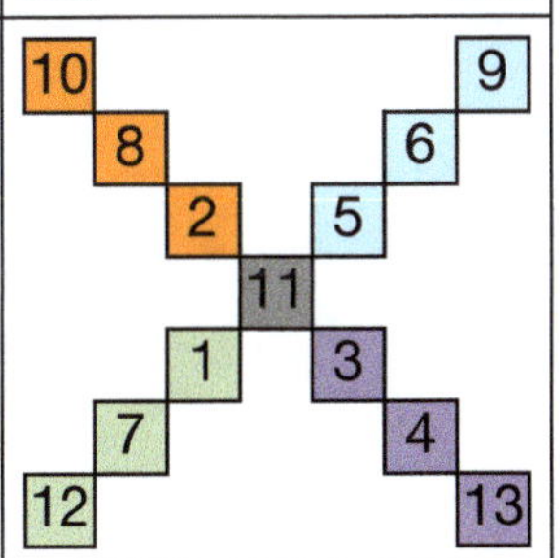

Some children may choose to investigate whether the X can be successfully completed using any 13 consecutive numbers (for example, 18–30). Any set of 13 consecutive numbers will work, and solutions can be found using these lettered guides. Write the numbers in order, then write a letter below each number in order from a–m. Then follow one of the guides to place the numbers in the X.

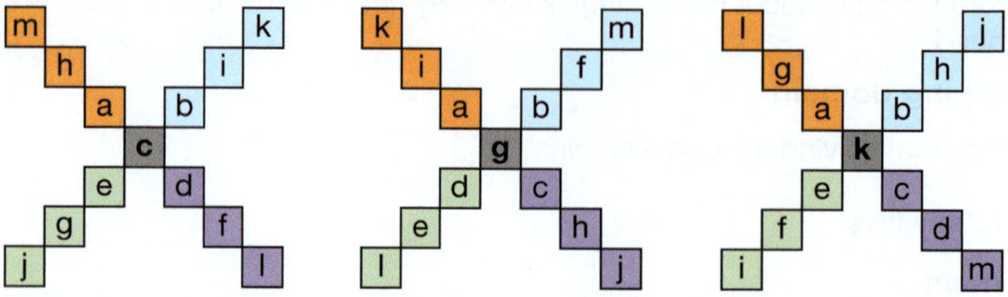

Some children may choose to investigate whether the X can be successfully completed using any 13 non-consecutive numbers. Depending on the numbers chosen, there may or may not be a solution.

A set of numbers that has a solution: 2, 5, 7, 8, 10, 14, 15, 16, 20, 21, 22, 24, 25

Total: 189

Centre number	Total for each arm (excluding centre number)	Total for each arm (including centre number)	Example
5	189 − 5 = 184 184 ÷ 4 = 46	46 + 5 = 51 *No solution*	With 21 as the centre number: 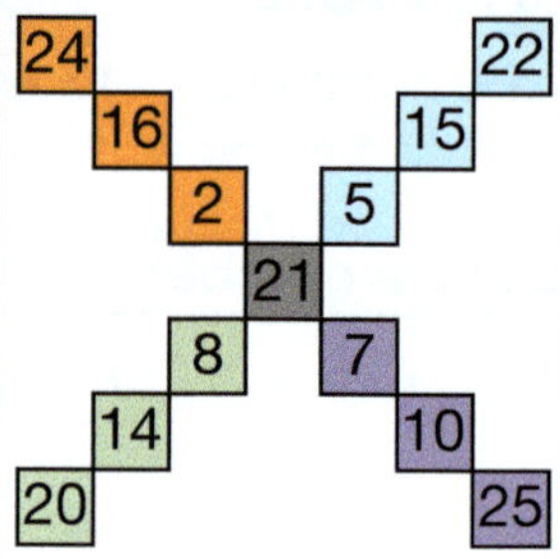
21	189 − 21 = 168 168 ÷ 4 = 42	42 + 21 = 63	
25	189 − 25 = 164 164 ÷ 4 = 41	41 + 25 = 66 *No solution*	

A set of numbers that does not work: 4, 11, 12, 20, 21, 22, 23, 25, 30, 32, 49, 51, 54

Total: 354

Centre number	Total for each arm (excluding centre number)	Total for each arm (including centre number)	Example
22	354 − 22 = 332 332 ÷ 4 = 83	83 + 22 = 105	*No solution*
30	354 − 30 = 324 324 ÷ 4 = 81	81 + 30 = 111	*No solution*
54	354 − 54 = 300 300 ÷ 4 = 75	75 + 54 = 129	*No solution*

Triple Trick

Year 3

Introduction

In **Triple Trick**, children will discover that the sum of three consecutive numbers can be found by multiplying the middle number by 3. This is because if you take 1 from the largest number and give it to the smallest number, all the numbers become the same as the middle number. For example, $4 + 5 + 6$ could be rebalanced to $5 + 5 + 5$, and the total can be found by calculating $3 \times 5 = 15$. Children need to understand conservation of number in order to see why this 'trick' works.

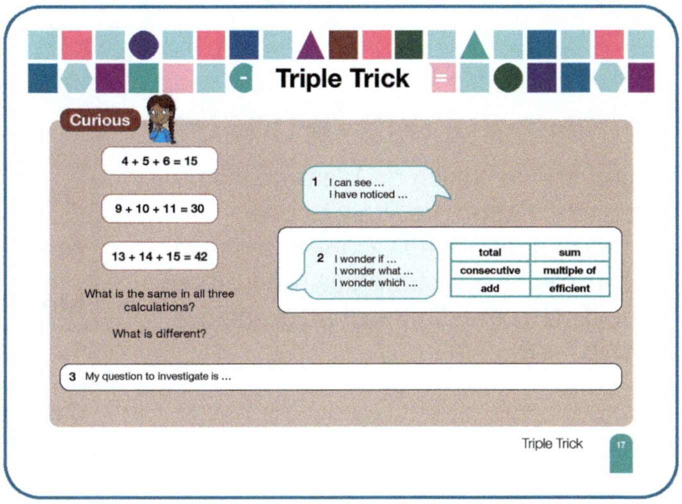

This task exemplifies the power of making a model. The relationship between the numbers, and the operations performed on them, becomes clear when children reveal the mathematical structures using concrete resources.

This task promotes greater-depth learning in a range of ways, supporting children to identify and articulate underlying mathematical structures. For example, in the **Communicate** stage, greater depth is promoted by asking children to develop clear, accurate and well-organised communication, using correct mathematical language and illustrating key ideas using diagrams or models.

This **Teacher Guide** is supported by photocopiable **Pupil Resources** and **Teacher's Presentation** PowerPoint slides for whole-class or group display.

National Curriculum

Triple Trick covers these Year 3 National Curriculum objectives.

Addition and subtraction

- Solve problems, including missing number problems, using number facts, place value, and more complex addition and subtraction.

Multiplication and division

- Recall and use multiplication and division facts for the 3, 4 and 8 multiplication tables.

The 6C cycle

The grid on the right shows the relative weighting given to each of the 6 Cs in **Triple Trick**.

Curious	★★
Connect	★★
Collaborate	★
Communicate	★★★
Create	★★
Consider	★

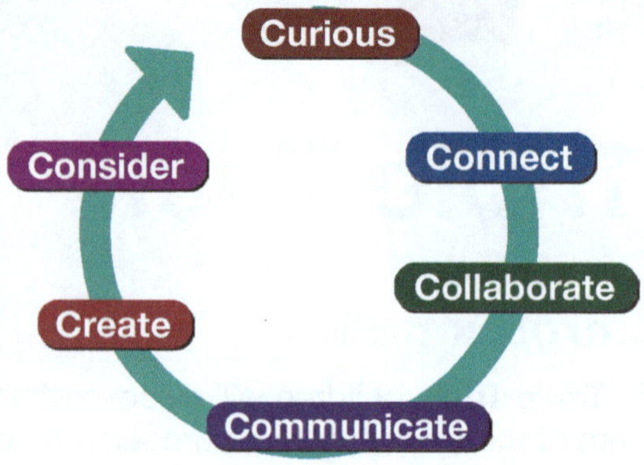

Problem-solving heuristics

The grid shows which problem-solving strategies children are likely to use when investigating the **Connect** tasks in **Triple Trick**. These strategies are further exemplified in the **Connect** stage below (pages 67–68).

Draw a diagram	
Make a table / spot a pattern	✓
Make a model / act it out	✓
Try then improve	
Create a systematic list	
Solve a simpler but related problem	
Work backwards	✓

Key mathematical ideas

The 'trick' is best exemplified using a model. For example, for the addition $3 + 4 + 5$, three towers of interlocking cubes could be built, using a different colour for each tower.

Moving one cube from the tallest tower to the shortest makes all the towers an equal height.

The towers are now all the same height as the middle tower, so the sum of three consecutive numbers can be found by multiplying the middle number by 3.

There is no expectation for children to respond using algebraic notation within the Year 3 or 4 National Curriculum. However, some children may find it interesting to explore how the notation matches what they have physically experienced, demystifying the symbols.

$$(n - 1) + n + (n + 1) = 3n \text{ (where } n \text{ is the middle number)}$$

When children move to the **Create** stage, the opportunities open up as this principle works for adding any odd series of consecutive numbers. For example, when adding five consecutive numbers, the total is 5 times the middle number.

$9 + 10 + 11 + 12 + 13 = 11 \times 5 = 55$ $\quad$ $38 + 39 + 40 + 41 + 42 = 40 \times 5 = 200$

Children working at greater depth should be intrigued to explore if the same rule applies to an even series of consecutive numbers. If not, can they find a rule that does work?

When adding four consecutive numbers, the total is double the sum of the two middle numbers.

$2 + 3 + 4 + 5 = 2 \times (3 + 4) = 2 \times 7 = 14$ $\quad$ $49 + 50 + 51 + 52 = 2 \times (50 + 51) = 2 \times 101 = 202$

Children should be able to see this rule works because the sum of the middle two numbers is equal to the sum of the first and last numbers (for example, for $2 + 3 + 4 + 5$, $2 + 5 = 3 + 4$).

Another rule that could be applied when adding four consecutive numbers is to multiply the first number by 4 and add 3.

Answers to the tasks are given on page 71.

Teaching guidance

If you are using the **Express Pathway**, please turn to pages 67–70.

Teaching guidance for the **Full 6C Pathway** is given below.

The **Teacher's Presentation** PowerPoint file has slides specifically mentioned in this teaching guidance. Some of the slides include notes (within the PowerPoint file) providing 'stumble support' suggestions, that may help you to support children in their problem-solving thinking and enable access for children who are less confident.

Awe and wonder

Slide 1

Display slide 1.

- *I can solve a calculation before I finish writing it.*
- *Give me a number less than 20.*

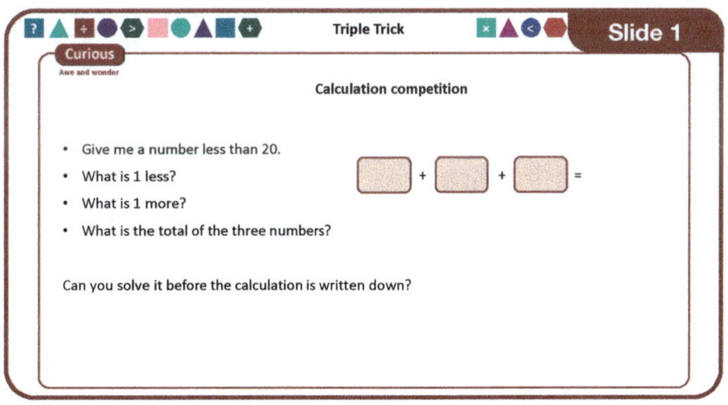

Write the number in the middle box, for example, 13.

- *Tell me the numbers 1 less and 1 more.*

Before writing these, say:

- *I already know the total.*

Write the total after the = symbol. (Multiply the middle number by 3, for example, $13 \times 3 = 39$.) Write the numbers 1 less and 1 more in the first and third boxes to complete the addition, for example, $12 + 13 + 14$.

Ask children to check your answer.

Slide 2

Display slide 2.

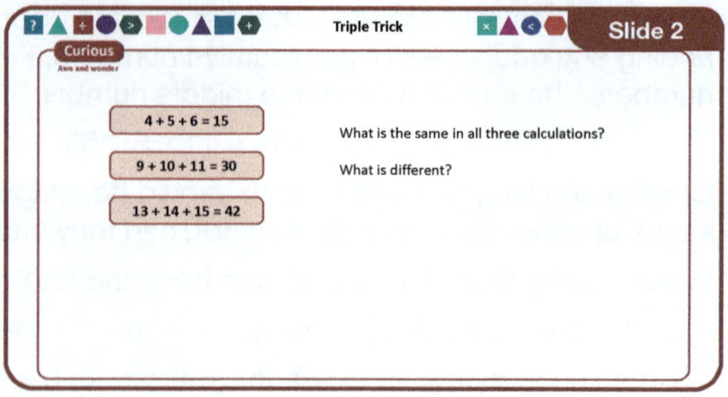

- I can see that each calculation has three numbers being added.
- I have noticed that one of the totals is an odd number.
- I have noticed that two of the totals are even numbers.
- I have noticed that the second total is double the first total.
- I can see that the numbers being added together are consecutive.
- I have noticed that all the totals are in the 3 times table.
- I can see that all the totals are two-digit numbers.
- I have noticed that each number is only used once.

Ensure that all children understand, and are comfortable using, the term 'consecutive'.

Develop and deepen

Slide 3

Display slide 3.

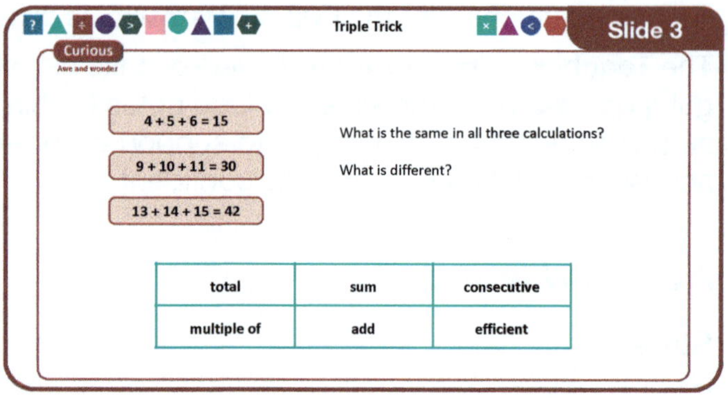

Encourage children to use the words in the word bank to develop and deepen their thoughts and questions.

total	**sum**	**consecutive**
multiple of	**add**	**efficient**

Possible lines of enquiry linked to the word bank:

- I wonder if the multiples of 3 repeat 'odd, even, odd, even …'
- I wonder if there is an efficient way of adding consecutive numbers.
- I wonder if the sum of three consecutive numbers is always a multiple of 3.
- I wonder which numbers in the 3 times table I can make by adding together three consecutive numbers.

Possible lines of enquiry that children may raise now, but should be held back until the **Create** task:

- I wonder what happens when four or five consecutive numbers are added together.

Consider whether / how to collate and display all the suggested questions, and whether / how to steer children towards selecting an appropriately challenging question.

Pupil Resources for Full and Express Pathways

Distribute the **Pupil Resources** for **Triple Trick** to the children to help support them in their exploration of the different tasks. Use either of the **Curious Pupil Resources** depending on whether you are following the **Full Pathway** or the **Express Pathway**. The **Pupil Resources** for the remaining 5Cs are common to both pathways.

To provide additional flexibility in the classroom, each of the **Pupil Resources** is also provided in the **Teacher's Presentation** PowerPoint file (see slides 4–7), so you can display these at any time if you wish.

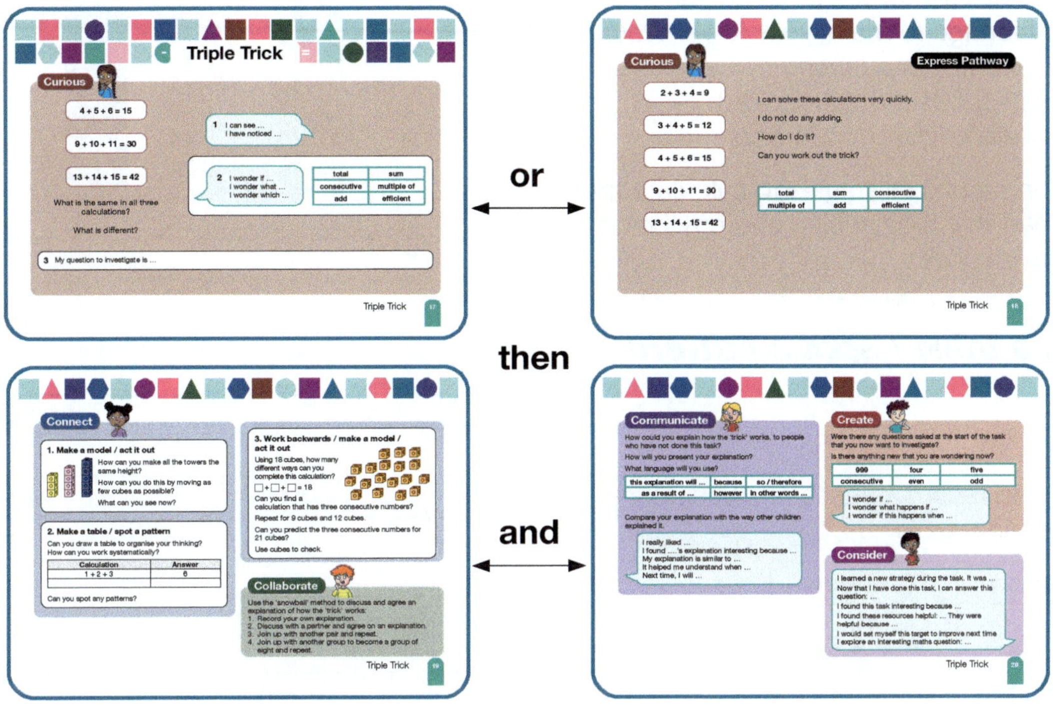

Express Pathway

If you are following the **Express Pathway**, spend a few minutes discussing the **Curious Pupil Resource** with the children. (This is slide 5, so you could display it if you wish.) Use some of the prompts on the previous page to help.

Three tasks involving different strategies are suggested for children to make connections.

1. Make a model / act it out

Additional resources required: interlocking cubes, geoboards and pegs, or Numicon Shapes

The easiest way to understand how the numbers are moving in this problem is to build it using concrete resources.

Encourage children to use a different colour for each number, to make it easier for them to spot the pattern.

Slide 8

Display slide 8.

- *What calculation does this represent?* (3 + 4 + 5)
- *What do you notice?* (The towers get 1 cube higher each time.)
- *Can you use the word 'consecutive'?*

Ask children to build the towers themselves.

- *Now can you make all the towers the same height?*

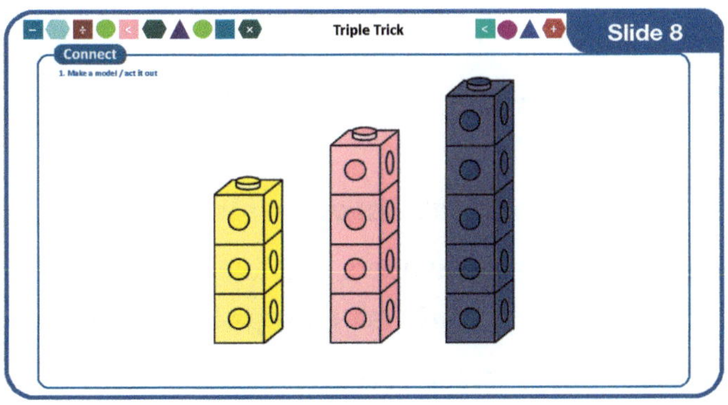

Slide 9

Display slide 9.

- *What can you see now? What has changed?*
- *Did you manage to make your towers look like this (all the same height)?*
- *How many cubes did you move?*
- *Try again with different numbers.*

A discussion around strategies to calculate the sum may lead to a rebalancing strategy being identified as efficient.

2. Make a table / spot a pattern

Slide 10

Display slide 10.

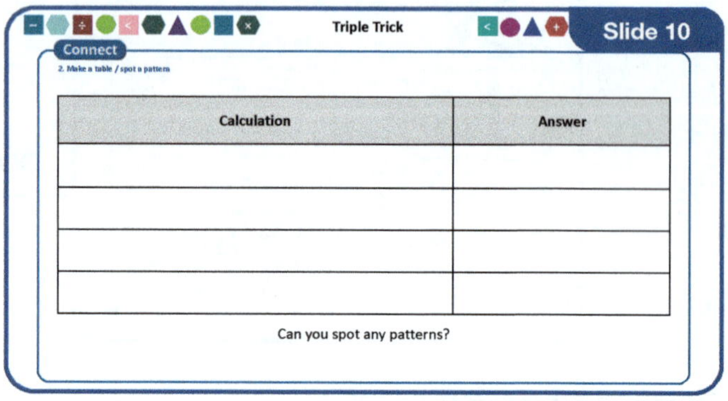

Children could record a series of calculations and answers in a table.

Calculation	Answer
1 + 2 + 3	6
2 + 3 + 4	9
3 + 4 + 5	12
4 + 5 + 6	15

- *Can you spot a pattern between the middle number in the calculation and the answer?*
- *Can you explain why this happens?*
- I know that to add three consecutive numbers together quickly you can multiply the middle number by 3. This is because if you take 1 from the largest number and give it to the smallest number, all the numbers become the same as the middle number. I can explain it using a diagram / table.
- *Is this always true?*
- *Can you prove it?*
- *Can you spot any other patterns?* (The answer is the same as the first number multiplied by 3, with another 3 added on.)

3. Work backwards / make a model / act it out

Additional resources required: interlocking cubes

Give children 18 cubes.

- *Using the cubes, how many different ways can you complete this calculation?*

$$\square + \square + \square = 18$$

- *Can you find a calculation that has three consecutive numbers?*

Repeat with other numbers of cubes (multiples of 3, such as 9 and 12).

Children could record their findings in a table.

- *What patterns can you see?*

Once children have identified a pattern, give them 21 cubes. Can they predict what the three consecutive numbers will be, and then prove it using the cubes?

Refer to the **Collaborate** task of the **Pupil Resource**.

Children could collaborate using the 'snowball' method. First, children each record their explanation of why the 'trick' works. They then work in pairs, sharing their explanations with each other, asking questions and discussing. They agree a joint 'answer' and jot it down. Pairs then form groups of four and repeat the process. And then groups of eight and repeat.

- *What patterns did you spot?*
- *Can you explain why this happens?*
- *How could you explain or show your findings as clearly as possible, so someone else understands the rule you have found?*

Refer to the **Communicate** task of the **Pupil Resource**.

This task provides an opportunity for children to perform the 'trick' for another class, or the whole school in assembly, or perhaps for parents/carers.

This would also be an excellent opportunity for children to develop a digital resource to explain how the 'trick' works. Due to the simplicity of the concrete resources involved, a stop motion animation would be possible. Alternatively, children could create an explanatory video for the school website or social media.

Slide 11

Display slide 11.

this explanation will …	**because**	**so / therefore**
as a result of …	**however**	**in other words …**

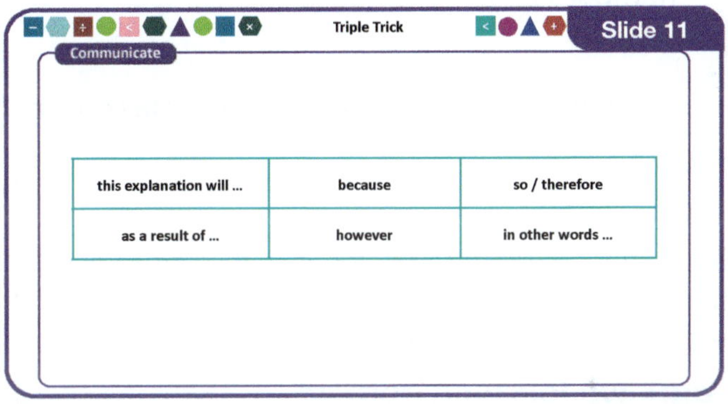

Slide 12

Display slide 12.

Afterwards, encourage children to compare their explanation with those of other children.

- I really liked …
- I found …'s explanation interesting because …
- My explanation is similar to …
- It helped me understand when …
- Next time I need to explain a mathematical pattern or rule, I will …

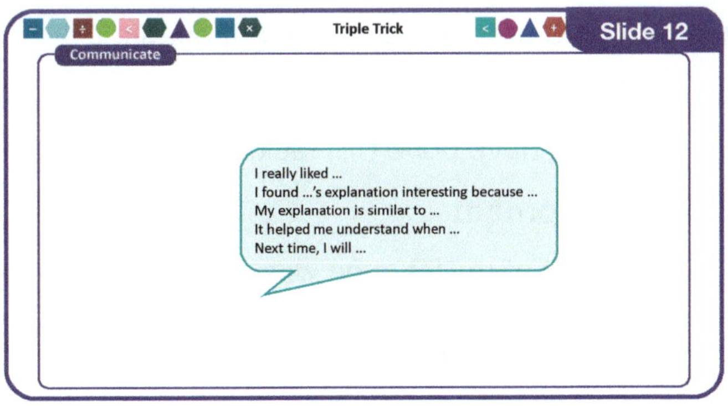

Create

Refer to the **Create** task of the **Pupil Resource**.

Slide 13

Display slide 13.

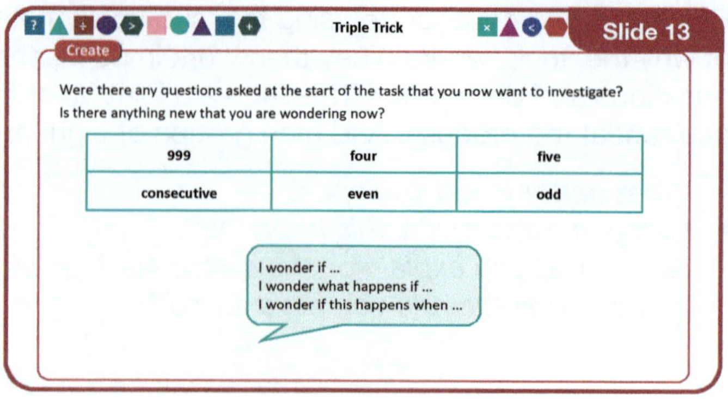

- *Were there any questions asked at the start of the task that you now want to investigate?*
- *Is there anything new that you are wondering now?*

Children can use the word bank to prompt further questions of their own.

999	**four**	**five**
consecutive	**even**	**odd**

- I wonder what happens when four or five consecutive numbers are added together.
- I wonder if I can find the three consecutive numbers that total 999.
- I wonder if this happens when I add any odd number of consecutive numbers.

⚠ Watch out for children who think that to add four consecutive numbers you multiply one of the numbers by 4.

Consider

Refer to the **Consider** task of the **Pupil Resource**.

Slide 14

Display slide 14.

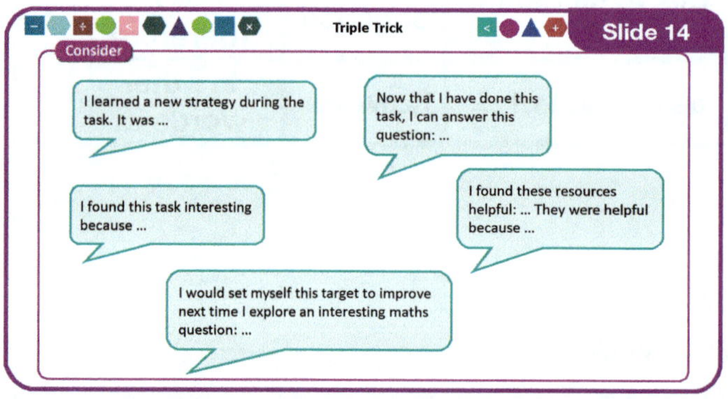

Encourage children to think about the strategies that they have used. The prompts below may help get children started.

Cognitive learning domain

- I learned a new strategy during the task. It was …
- The trickiest bit was …
- Now that I have done this task, I can answer this question: …

Affective domain

- The fun bit of my learning was … It was fun because …
- I was surprised about … because …
- I found this task interesting because …

Metacognitive domain

- I got stuck when … I unstuck myself by …
- I found these resources helpful: … They were helpful because …
- I would set myself this target to improve next time I explore an interesting maths question: …

Answers

Number of consecutive numbers	Rule to find the total	Expanded rule	Example
Three (odd)	Where n is the middle number: **$3n$**	$(n-1)+n+(n+1)$ $=3n-1+1$ $=3n$	$332+333+334$ $=333\times3=999$
	OR: Where f is the first number: **$3f+3$**	$f+(f+1)+(f+2)$ $=3f+1+2$ $=3f+3$	$332+333+334$ $=(332\times3)+3$ $=996+3$ $=999$
Five (odd)	Where n is the middle number: **$5n$**	$(n-2)+(n-1)+n+(n+1)+(n+2)$ $=5n-2-1+1+2$ $=5n$	$10+11+12+13+14$ $=12\times5=60$
	OR: Where f is the first number: **$5f+10$**	$f+(f+1)+(f+2)+(f+3)+(f+4)$ $=5f+1+2+3+4$ $=5f+10$	$10+11+12+13+14$ $=(10\times5)+10$ $=50+10$ $=60$
Four (even)	Where p and q are the middle two numbers: **$2(p+q)$**	$(p-1)+p+q+(q+1)$ $=2p+2q-1+1$ $=2p+2q$ $=2(p+q)$	$19+20+21+22$ $=(20+21)\times2$ $=41\times2$ $=82$
	OR: Where f is the first number: **$4f+6$**	$f+(f+1)+(f+2)+(f+3)$ $=4f+1+2+3$ $=4f+6$	$19+20+21+22$ $=(19\times4)+6$ $=76+6$ $=82$
Six (even)	Where p and q are the middle two numbers: **$3(p+q)$**	$(p-2)+(p-1)+p+q+(q+1)+(q+2)$ $=3p+3q-2-1+1+2$ $=3p+3q$ $=3(p+q)$	$9+10+11+12+13$ $+14$ $=(11+12)\times3$ $=69$
	OR: Where f is the first number: **$6f+15$**	$f+(f+1)+(f+2)+(f+3)+(f+4)+(f+5)$ $=6f+1+2+3+4+5$ $=6f+15$	$9+10+11+12+13$ $+14$ $=(9\times6)+15$ $=54+15$ $=69$

Marching Elephants

Year 3

Year 4

Introduction

Marching Elephants requires children to use multiplication and division facts to find a mystery number. Children investigate the formations (arrays) a group of elephants form as they march, with remainders, to work out the total number of elephants. The task prompts children to use reasoning to work logically through the problem, rather than trying out every possible total.

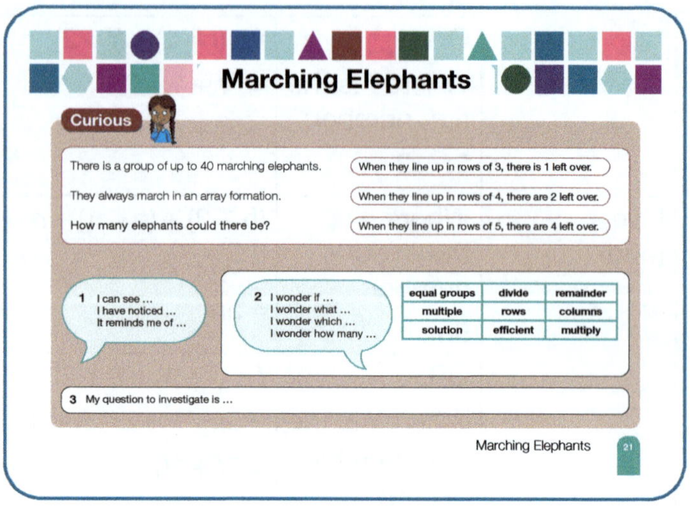

The children begin by creating arrays with equal rows of different numbers to represent the elephants. They use a systematic table or a model to record the possible totals until they identify one that satisfies all the clues. They can then demonstrate their solutions by creating a presentation.

Children apply and deepen their understanding by writing their own puzzles. They could be challenged by working within particular parameters, for example, the number of solutions and / or clues they may use.

This task promotes greater-depth learning in a range of ways, supporting children to identify and articulate underlying mathematical structures. For example, in the **Create** stage, greater depth is promoted by asking children to form their own mathematically accurate questions that are creative and inventive.

This **Teacher Guide** is supported by photocopiable **Pupil Resources** and **Teacher's Presentation** PowerPoint slides for whole-class or group display.

National Curriculum

Marching Elephants covers these Year 3 and Year 4 National Curriculum objectives.

Year 3 Multiplication and division

- Recall and use multiplication and division facts for the 3, 4 and 8 multiplication tables.

Year 4 Multiplication and division

- Recall multiplication and division facts for multiplication tables up to 12 × 12.
- Use place value, known and derived facts to multiply and divide mentally.
- Solve problems involving multiplying.

The 6C cycle

The grid below shows the relative weighting given to each of the 6 Cs in **Marching Elephants.**

Curious	★★
Connect	★★
Collaborate	★
Communicate	★
Create	★★★
Consider	★★

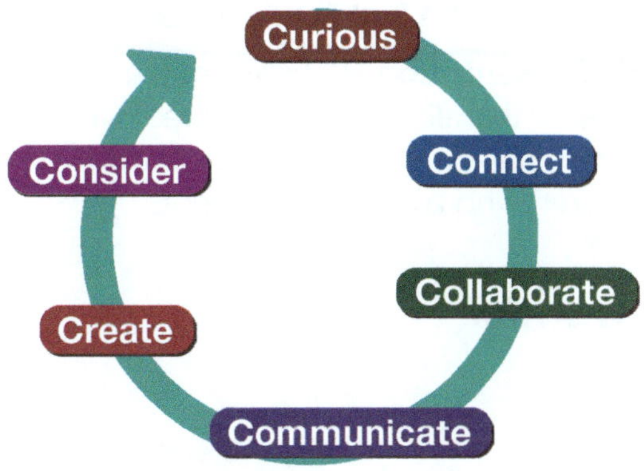

Problem-solving heuristics

The grid shows which problem-solving strategies children are likely to use when investigating the Connect tasks in **Marching Elephants**. These strategies are further exemplified in the Connect stage below (pages 76–77).

Draw a diagram	✓
Make a table / spot a pattern	✓
Make a model / act it out	
Try then improve	
Create a systematic list	✓
Solve a simpler but related problem	
Work backwards	

Key mathematical ideas

The aim of this task is for children to understand the efficiency and greater reliability of working strategically rather than using a trial-and-error method.

They are challenged to assess the relative importance and relevance of clues to a mystery number, using them to eliminate impossible options. In the interests of efficiency, children must identify which clues will eliminate the most numbers and follow the others to refine their options.

In its focus on arrays, the task provides the opportunity for children to relate concrete resources to abstract mathematics. It develops children's ability to apply their knowledge of division and remainders in a narrative context. Children's known multiplication and division facts will also be tested when they are prompted to recognise and work from multiples, and their related facts, quickly. For example:

- If there are some elephants left over when they line up in rows of either 3 or 5, the number of elephants cannot be a multiple of 3 or 5.
- If there are 2 elephants left over when they line up in rows of 4, the total is a multiple of 2 but not a multiple of 4.

Children will additionally be prompted to develop their ability to provide mathematical proof. They are required to record their investigative approaches and results clearly, and finally to present their methods and solutions in a demonstration.

Answers to the tasks are given in situ, and also summarised on page 80.

Teaching guidance

If you are using the **Express Pathway**, please turn to pages 76–79.

Teaching guidance for the **Full 6C Pathway** is given below.

The **Teacher's Presentation** PowerPoint file has slides specifically mentioned in this teaching guidance. Some of the slides include notes (within the PowerPoint file) providing 'stumble support' suggestions, that may help you to support children in their problem-solving thinking and enable access for children who are less confident.

Awe and wonder

Slides 1–4

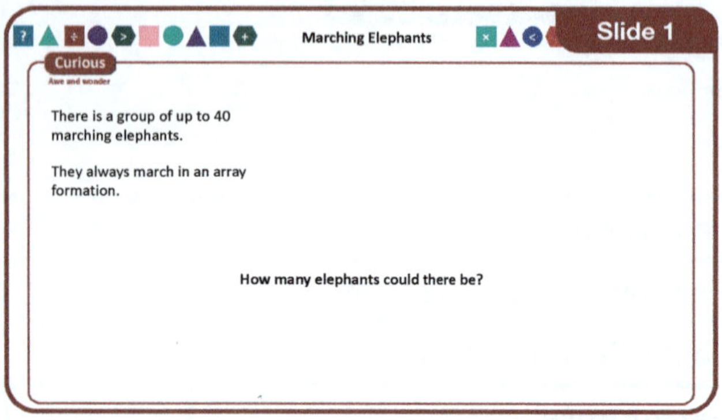

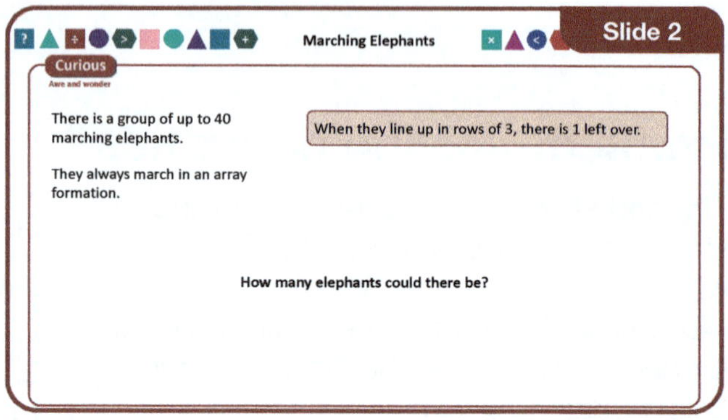

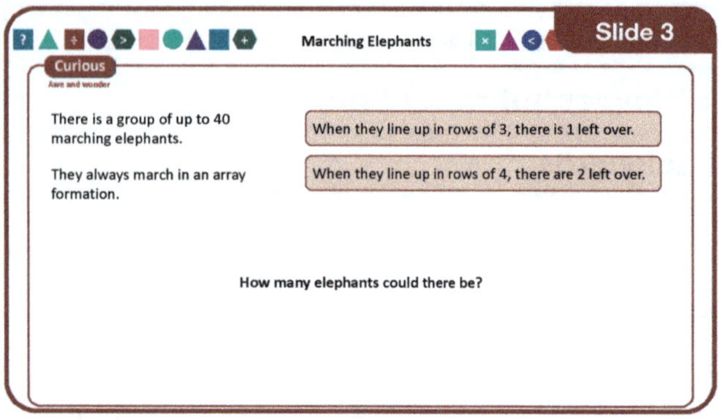

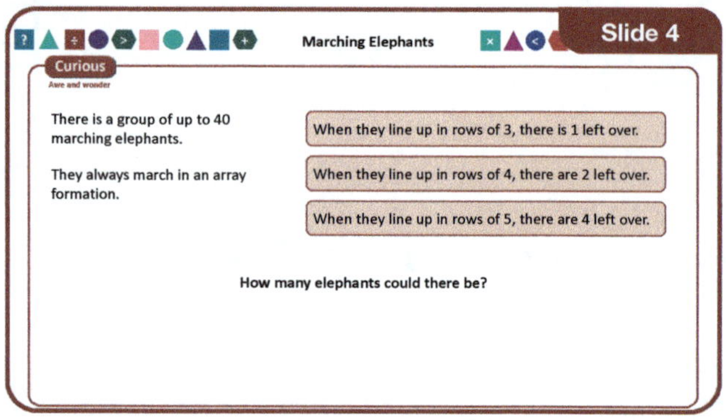

You may choose to reveal the information line by line to aid discussion and observations as prompted below (displaying slides 1–4 in turn), or all at once (displaying only slide 4).

Encourage children to share what they are noticing or wondering.

- I can see that there are at least 9 elephants: when they are in groups of 5, there are 4 left over.
- I have noticed that the number of elephants cannot be a multiple of 3, 4 or 5 because there is a remainder when they are divided into these group sizes.
- I know that if the elephants are marching in an array, it must be a rectangle.
- This reminds me of lining up like soldiers.
- I know that there can't be 54 elephants because the total number is up to 40.
- I know that there will be some remainders when the elephants are divided into groups of 3, 4 or 5.

Develop and deepen

Slide 5

Display slide 5.

Encourage children to use the words in the word bank to develop and deepen their thoughts and questions.

equal groups	**divide**	**remainder**
multiple	**rows**	**columns**
solution	**efficient**	**multiply**

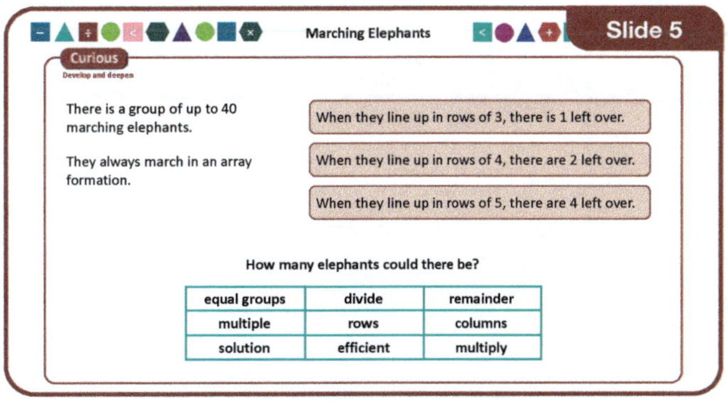

Possible lines of enquiry:

- I wonder whether the total number of elephants is going to be odd or even.
- I wonder how many elephants there could be.
- I wonder what is the smallest possible number of elephants.

Possible lines of enquiry linked to the word bank:

- I wonder what would be the most efficient way of finding out how many elephants there are.
- I wonder if there is more than one possible solution.

Possible lines of enquiry that children may raise now, but should be held back until the **Create** task:

- I wonder if I could write my own number riddle.

Consider whether / how to collate and display all the suggested questions, and whether / how to steer children towards selecting an appropriately challenging question.

Pupil Resources for Full and Express Pathways

Distribute the **Pupil Resources** for **Marching Elephants** to the children to help support them in their exploration of the different tasks. Use either of the **Curious Pupil Resources** depending on whether you are following the **Full Pathway** or the **Express Pathway**. The **Pupil Resources** for the remaining 5Cs are common to both pathways.

To provide additional flexibility in the classroom, each of the **Pupil Resources** is also provided in the **Teacher's Presentation** PowerPoint file (see slides 6–9), so you can display these at any time if you wish.

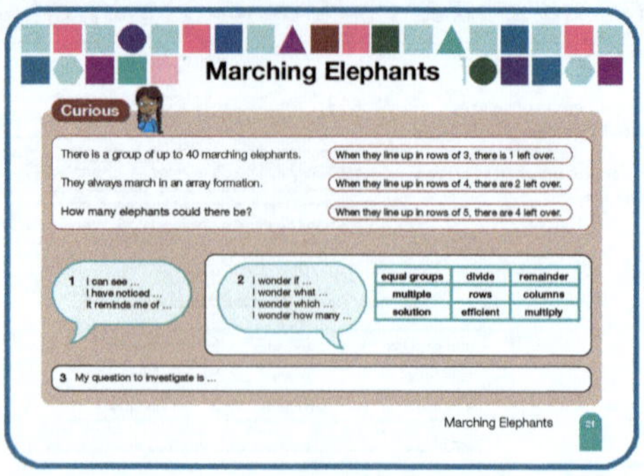

or

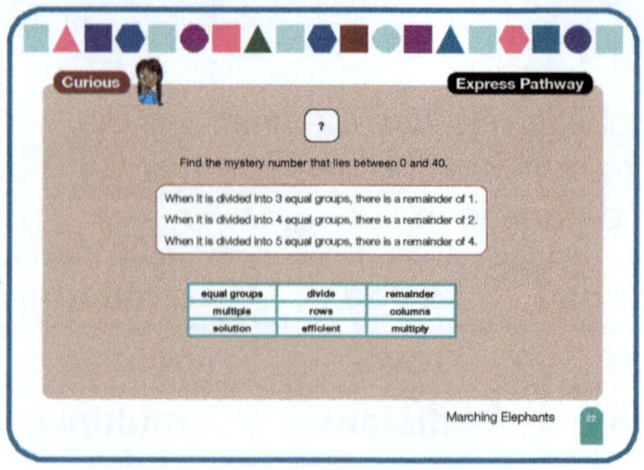

then

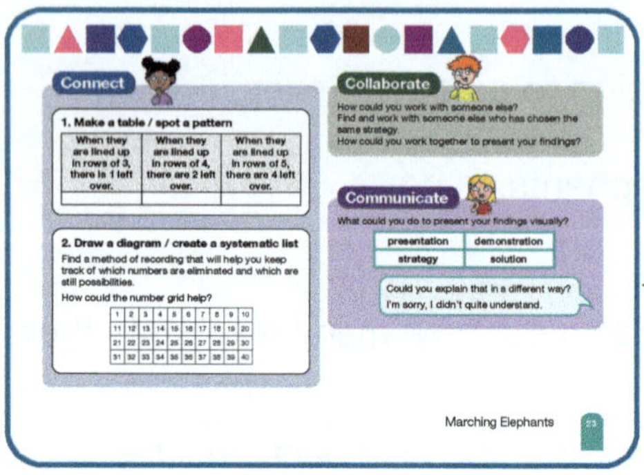

and

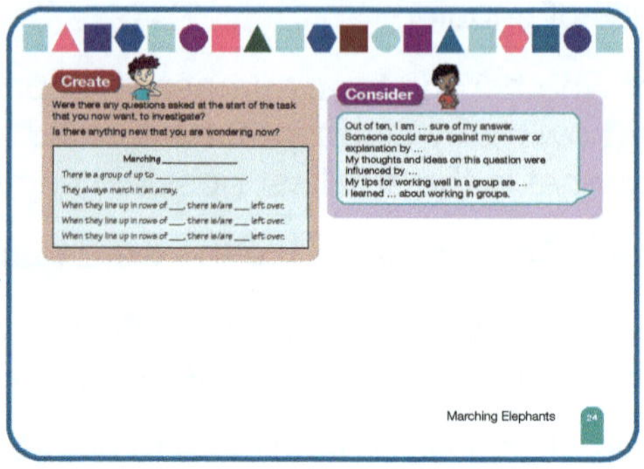

Express Pathway

If you are following the **Express Pathway**, spend a few minutes discussing the **Curious Pupil Resource** with the children. (This is slide 7, so you could display it if you wish.) Use some of the prompts on the previous page to help.

Two tasks involving different strategies are suggested for children to make connections.

1. Make a table / spot a pattern

Additional resources required: cubes or counters (optional)

Slide 10

Print slide 10 and distribute to children.

Suggest one approach to solving the problem: children could use the table to list the numbers that satisfy each of the clues, and then find which number occurs in all three lists.

It is likely that children will not start systematically, so they may need guidance.

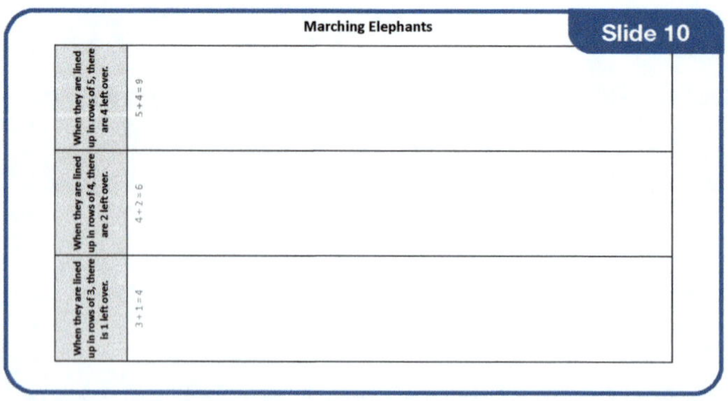

When they are lined up in rows of 3, there is 1 left over.	When they are lined up in rows of 4, there are 2 left over.	When they are lined up in rows of 5, there are 4 left over.
3 + 1 = 4	4 + 2 = 6	5 + 4 = 9
6 + 1 = 7	8 + 2 = 10	10 + 4 =14
9 + 1 = 10	12 + 2 = 14	15 + 4 = 19
12 + 1 = 13	16 + 2 = 18	20 + 4 = 24
15 + 1 = 16	20 + 2 = 22	25 + 4 = 29
18 + 1 = 19	24 + 2 = 26	30 + 4 = 34
21 + 1 = 22	28 + 2 = 30	35 + 4 = 39
24 + 1 = 25	32 + 2 = 34	
27 + 1 = 28	36 + 2 = 38	
30 + 1 = 31		
33 + 1 = 34		
36 + 1 = 37		

⚠ Watch out for children who do not seem to understand the mathematical structures. Encourage them to build the first few arrays using concrete resources such as cubes or counters. They could construct, for example, 2 rows of 3 with 1 left over.

2. Draw a diagram / create a systematic list

Slide 11

Print slide 11 and distribute to children.

This method is designed to remove possibilities efficiently, by discounting numbers that do not fit the clues. Children will need a method of recording to keep track of which numbers are eliminated and which remain as possibilities. One approach is to shade or colour numbers on the section of a 1–100 number square. For example:

Marching Elephants — Slide 11

1	2	3	4	5	6	7	8	9	10
11	12	13	14	15	16	17	18	19	20
21	22	23	24	25	26	27	28	29	30
31	32	33	34	35	36	37	38	39	40

- I know the total must be an even number because, when it is divided by 4, there are 2 left over. So I crossed out all the odd numbers using blue pencil.
- I know the total cannot be a multiple of 3, 4 or 5 because they each produced a remainder. So I crossed out all the remaining multiples of 3 using red pencil, then the remaining multiples of 4 using green pencil, and then the remaining multiples of 5 using purple pencil.
- I know the total is 1 more than a multiple of 3. So I crossed out the remaining numbers that are not 1 more than a multiple of 3, using orange pencil.

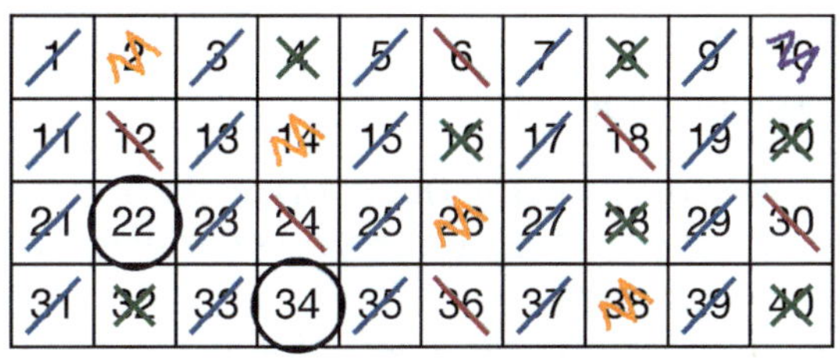

The above logic removes 38 possibilities, leaving only two numbers: 22 and 34. It is then possible to test a clue. In this example, children would look at which of the two leave a remainder of 4 when divided by 5. The solution of 34 is then revealed.

Collaborate

Refer to the **Collaborate** task of the **Pupil Resource**.

There are two main ways of collaborating during this task.

During the **Connect** stage, children could be paired up with those who have chosen the same strategy. For example, those choosing to record using a table could work together, and those working on systematically excluding possibilities using a number grid could form a different group.

Additionally, or alternatively, children could form pairs or small groups once the task has been solved, in preparation for the **Communicate** stage. Ask children to prepare to present the solution to the task visually, as a presentation or demonstration.

- *What strategies did you use?*
- *What was the same about how you worked?*
- *What was different?*
- *How could you explain your strategy and results to other children?*

Communicate

Refer to the **Communicate** task of the **Pupil Resource**.

Slide 12

Display slide 12.

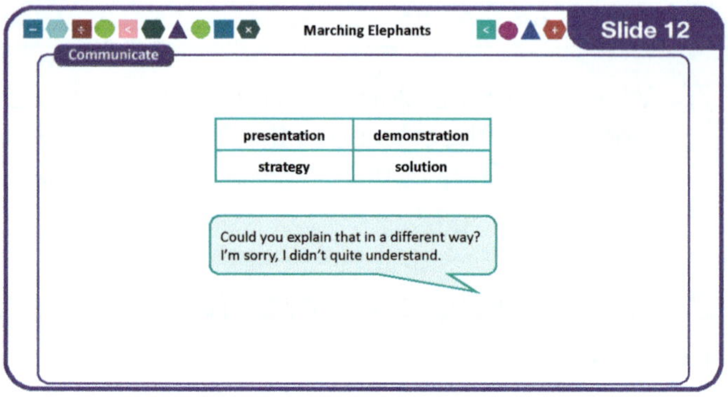

This task provides a good opportunity for children to present their findings visually. Their presentations could include live demonstrations and / or video clips of presentations or stop-motion animations.

Encourage children to use the word bank during their explanations.

presentation	**demonstration**
strategy	**solution**

Encourage them to ask questions during or after other children's presentations to clarify their ideas.

- Could you explain that in a different way?
- I'm sorry, I didn't quite understand.

Create

Refer to the **Create** task of the **Pupil Resource**.

Slide 13

Display slide 13.

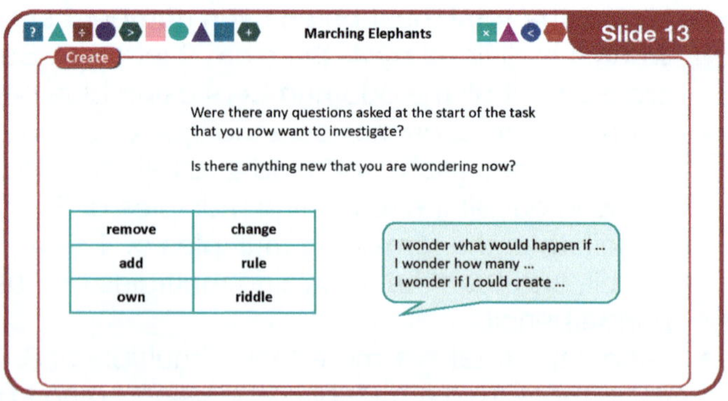

- *Were there any questions asked at the start of the task that you now want to investigate?*
- *Is there anything new that you are wondering now?*

Children can use the word bank to prompt further questions of their own.

remove	**change**
add	**rule**
own	**riddle**

- I wonder what would happen if the 1–40 rule was removed or changed.
- I wonder how many possible solutions there would be if I removed just one clue.
- I wonder if I could write my own number riddle.

Slide 14

Print slide 14 and distribute to children to support them in writing their own puzzles.

To create a further challenge, you could give parameters for children's puzzles. For example:

- *The riddle must have exactly ___ possible solutions.*
- *A maximum of ___ clues is allowed.*

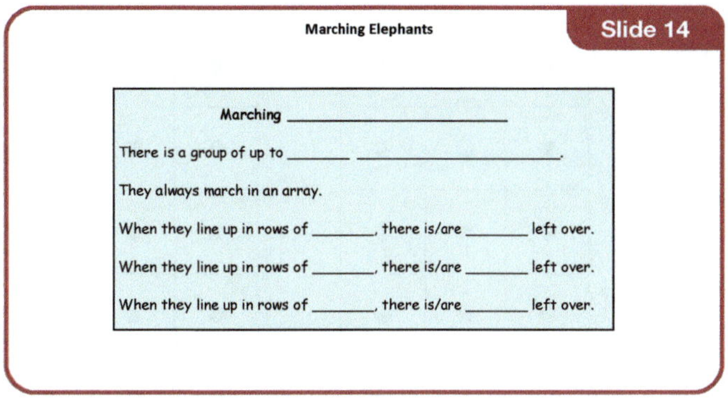

Refer to the **Consider** task of the **Pupil Resource**.

Slide 15

Display slide 15. Encourage children to think about the strategies they have used. The prompts below may help to get them started.

Cognitive learning domain

- The trickiest part of solving my question was …
- Out of ten, I am … sure of my answer.
- Someone could argue against my answer or explanation by …

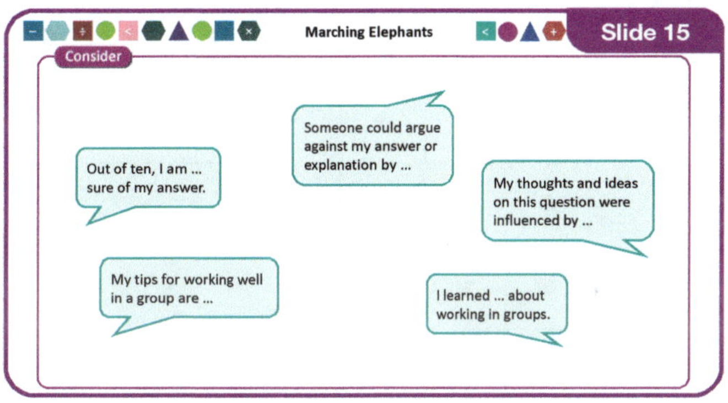

Affective domain

- The goal I will set myself next is …
- My thoughts and ideas on this question were influenced by ...
- I can feel more positive through my learning by …

Metacognitive domain

- I interacted with others by …
- My tips for working well in a group are ...
- I learned ... about working in groups.

Answers

There is only one solution to the prompt: 34 elephants. This is the only number between 0 and 40 that satisfies all three clues:

- When they line up in rows of 3, there is 1 left over.
- When they line up in rows of 4, there are 2 left over.
- When they line up in rows of 5, there are 4 left over.

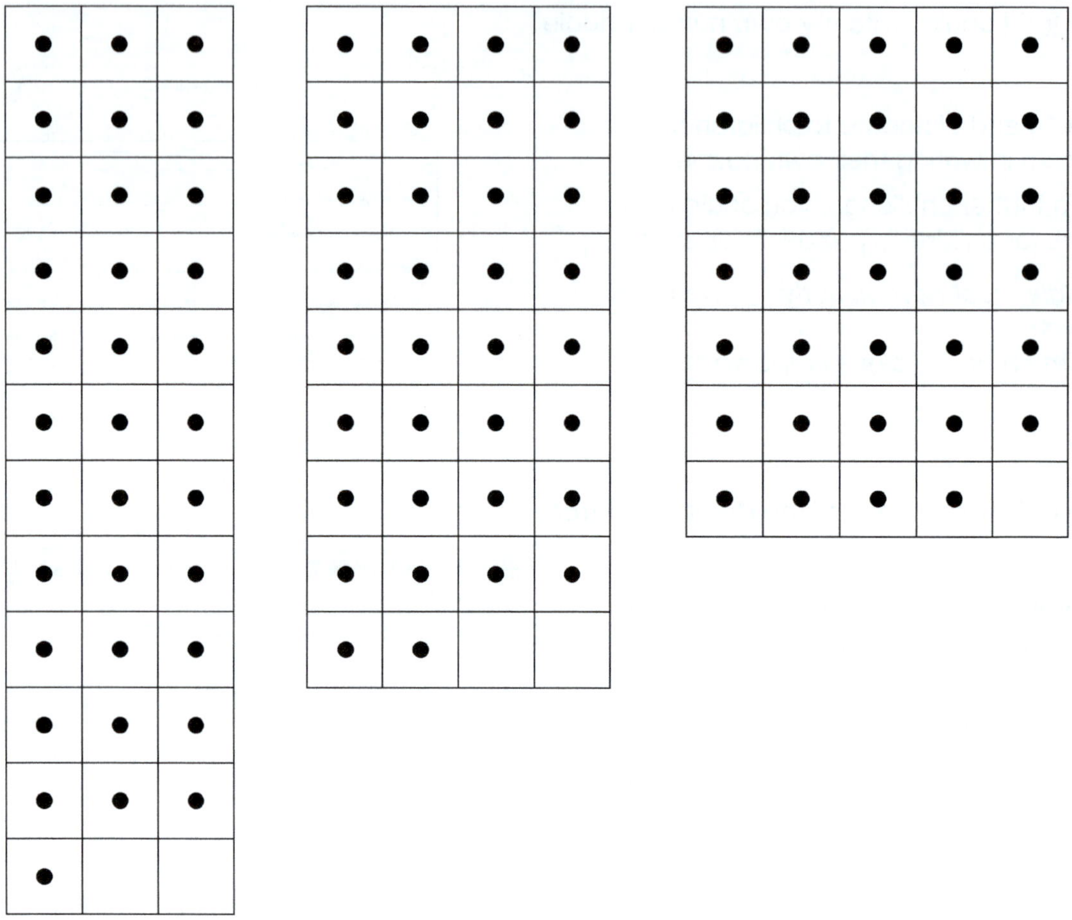

Fair and Square

Year 4

Introduction

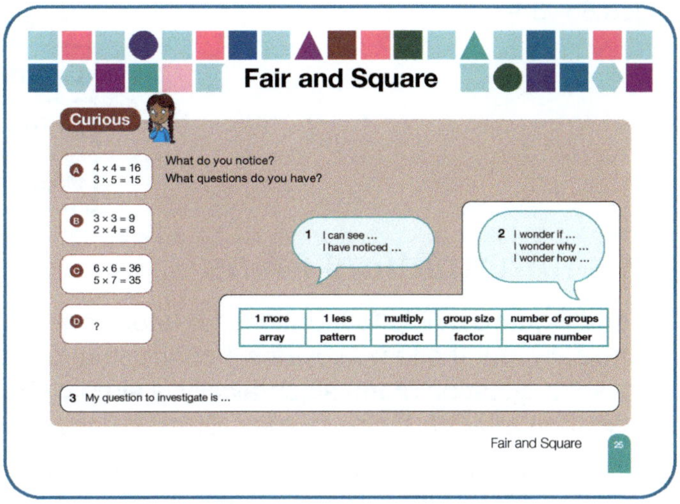

Fair and Square requires children to look deeply at the structure of, and relationships between, pairs of multiplications that form a pattern. The focus is on exploring why the pattern occurs (rather than on finding a numerical answer).

Some children may build a model to represent the similarities and differences between the calculations in each pair. Some may begin to spot patterns in the numbers being multiplied to find a product, and others may choose to work backwards from the product to find two pairs of numbers to multiply.

Children share and contrast different strategies to explain why the pattern occurs. They are encouraged to reflect critically on both their chosen strategy and the strategies presented by others.

This task promotes greater-depth learning in a range of ways, supporting children to identify and articulate underlying mathematical structures. For example, in the **Connect** stage, greater depth is promoted by asking children to explore a range of different mathematical resources and models, choose appropriate resources for the task, and explain their choice.

This **Teacher Guide** is supported by photocopiable **Pupil Resources** and **Teacher's Presentation** PowerPoint slides for whole-class or group display.

National Curriculum

Fair and Square covers these Year 4 National Curriculum objectives.

Multiplication and division

- Recall multiplication and division facts for multiplication tables up to 12×12.

Measurement

- Find the area of rectilinear shapes by counting squares.

The 6C cycle

The grid on the right shows the relative weighting given to each of the 6 Cs in **Fair and Square**.

Curious	★
Connect	★★★
Collaborate	★★
Communicate	★★
Create	★
Consider	★★

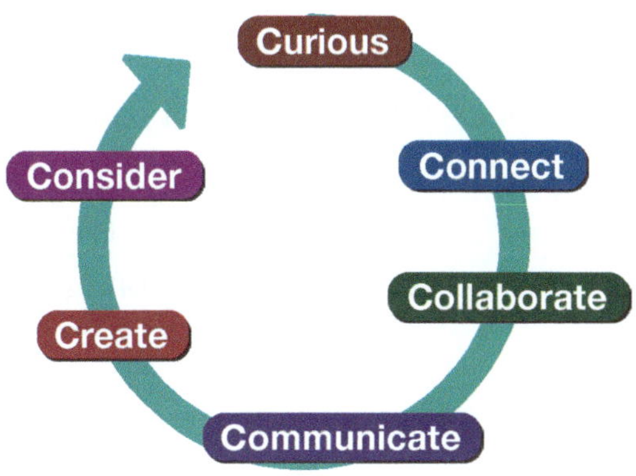

Problem-solving heuristics

The grid shows which problem-solving strategies children are likely to use when investigating the **Connect** tasks in **Fair and Square**. These strategies are further exemplified in the **Connect** stage below (pages 85–86).

Draw a diagram	
Make a table / spot a pattern	✓
Make a model / act it out	✓
Try and then improve	
Create a systematic list	
Solve a simpler but related problem	
Work backwards	✓

Key mathematical ideas

Fair and Square builds foundations for later work on algebra. Children are not expected to record algebraically, but the task provides a conceptual opportunity to experience that using letters (or, in this case, a question mark) algebraically is simply shorthand for a value that is yet to be found. This investigation also starts the journey towards understanding the difference between two squares.

Children are given a simple, easy-to-identify pattern to explore. In steps, they adjust the model, and therefore the calculation, to reduce the number of groups by 1 and increase the size of each group by 1. The challenge is in understanding and communicating why this happens.

Children may initially assume that the product remains the same in the new array, but this is not true. When working with concrete resources, children will rearrange the arrays to see what is happening, and identify the remainder in each case.

- Take 9 counters and arrange them into 3 rows of 3 (a 3 by 3 array). One row in the model is a different colour, so it can be tracked easily when it becomes part of the new array.
- Move 1 counter from the bottom row to the top row, to increase the number in the top row to 4.
- Repeat with the next counter from the bottom row. You will then have 1 counter left over that does not fit into the array.

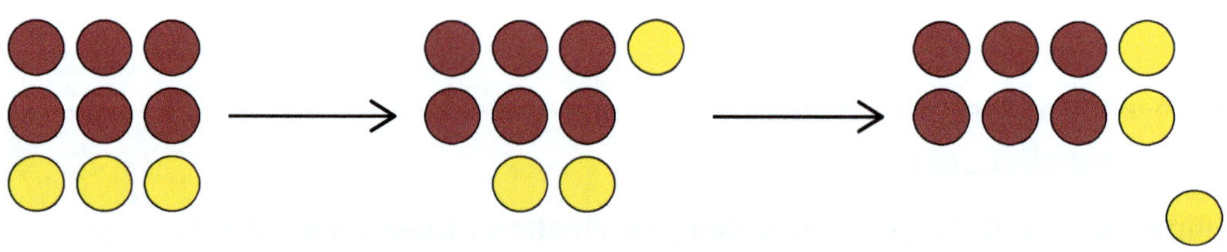

The new array contains 2 groups of 4 (a 2 by 4 array). The size of each group has been increased by 1 but there is 1 fewer group, which leaves 1 counter left over.

$$(3-1)\times(3+1)=$$

$$(3\times 3)+(3\times 1)-(1\times 3)-(1\times 1)=$$

$$9+3-3-1=8$$

If we replace the 3 with n, so it can be any number, then:

$$n^2 \text{ (square number)}$$

$$(n-1)(n+1)=n^2+n-n-1=n^2-1$$

Answers to the tasks are given on page 88.

Teaching guidance

If you are using the **Express Pathway**, please turn to pages 85–88.

Teaching guidance for the **Full 6C Pathway** is given below.

The **Teacher's Presentation** PowerPoint file has slides specifically mentioned in this teaching guidance. Some of the slides include notes (within the PowerPoint file) providing 'stumble support' suggestions, that may help you to support children in their problem-solving thinking and enable access for children who are less confident.

Awe and wonder

Slides 1–4

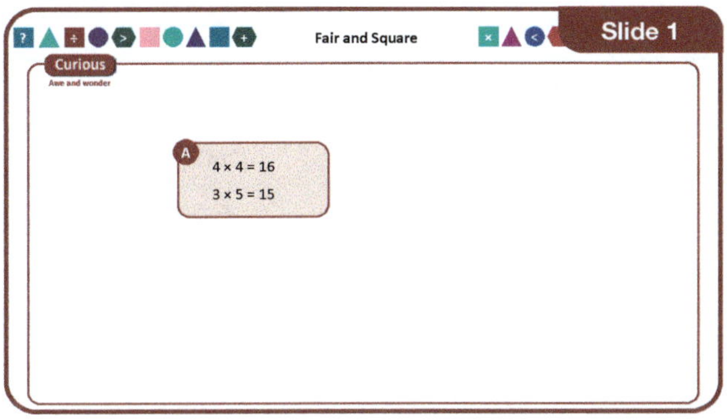

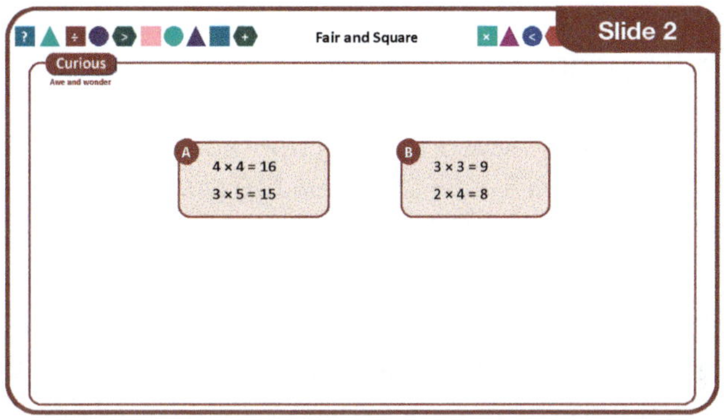

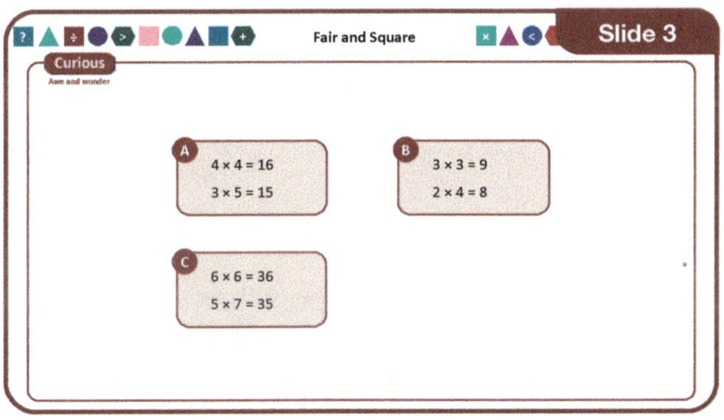

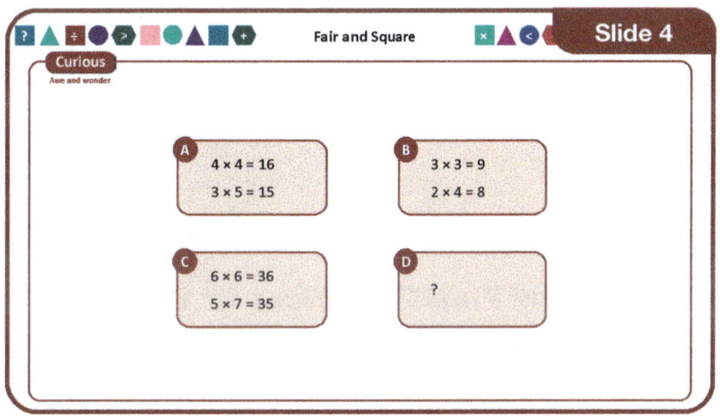

You may choose to display slides 1–4 one at a time, allowing time for discussion and observations as prompted below, or to reveal all of the multiplications at once (showing only slide 4). You may also wish to print slide 4 and distribute to children.

- *What is the same? What is different?*
- I have noticed that the second product is one less than the first product.
- I can see that they are all multiplications.
- I can see that they are in pairs.
- I can see that the first product in the pair is more than the second product.
- I can see that there are some square numbers.

For clue D, ask children to create their own pair of calculations, following the same pattern as clues A, B and C. Collect different examples from the children.

⚠ You may want to reinforce the word 'product' at this point and introduce the word 'factor'. Being able to use precise mathematical language will make it much easier for children to discuss the calculations.

Develop and deepen

Slide 5

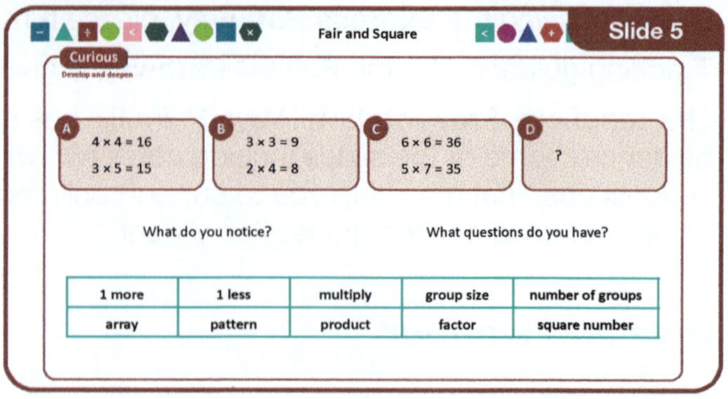

Display slide 5.

Encourage children to use words from the word bank to develop and deepen their thoughts and questions.

Possible lines of enquiry:

- I wonder if it is always true that odd × odd = odd.

Possible lines of enquiry linked to the word bank:

- I wonder why the second product is 1 less than the first product.
- I wonder how changing the factors by 1 more and 1 less changes the product.
- I wonder if this pattern always works for any pair of multiplications like this.

1 more	**array**
1 less	**pattern**
multiply	**product**
group size	**factor**
number of groups	**square number**

Possible lines of enquiry that children may raise now, but should be held back until the **Create** task:

- I wonder if it works with only … numbers.
- I wonder if it works with two-digit numbers.

Consider whether / how to collate and display all the suggested questions, and whether / how to steer children towards selecting an appropriately challenging question.

Pupil Resources for Full and Express Pathways

Distribute the **Pupil Resources** for **Fair and Square** to the children to help support them in their exploration of the different tasks. Use either of the **Curious** **Pupil Resources** depending on whether you are following the **Full Pathway** or the **Express Pathway**. The **Pupil Resources** for the remaining 5Cs are common to both pathways.

To provide additional flexibility in the classroom, each of the **Pupil Resources** is also provided in the **Teacher's Presentation** PowerPoint file (see slides 6–9), so you can display these at any time if you wish.

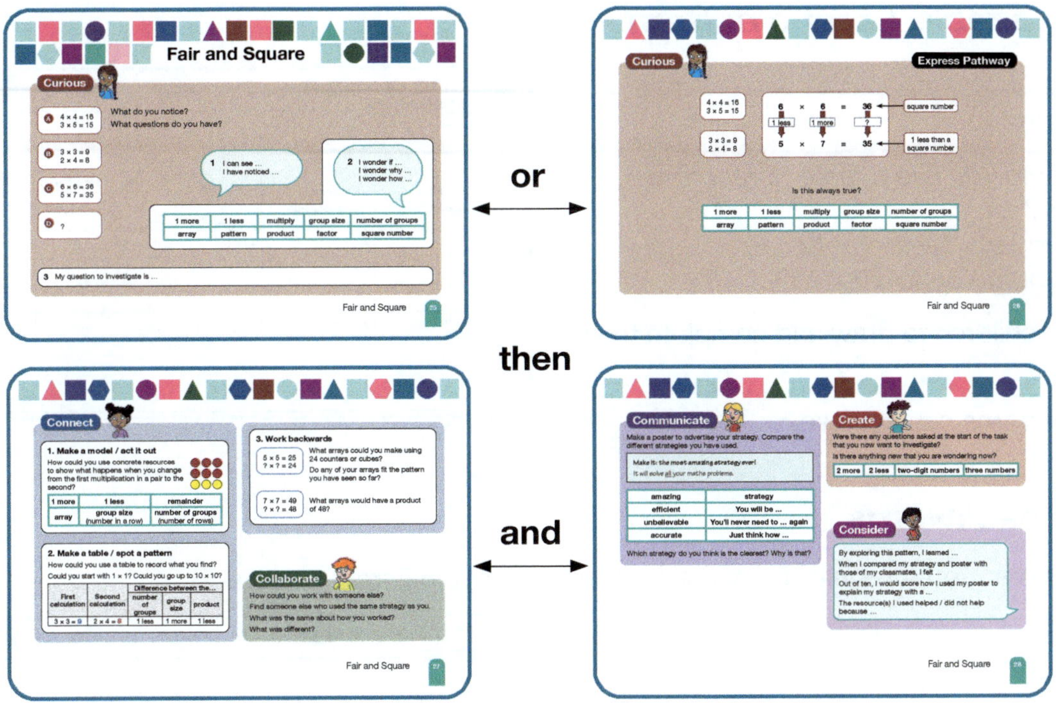

Express Pathway

If you are following the **Express Pathway**, spend a few minutes discussing the **Curious Pupil Resource** with the children. (This is slide 7, so you could display it if you wish.) Use some of the prompts on the previous page to help.

Three tasks involving different strategies are suggested for children to make connections.

1. Make a model / act it out

Additional resources required: counters or cubes in two different colours

Slide 10

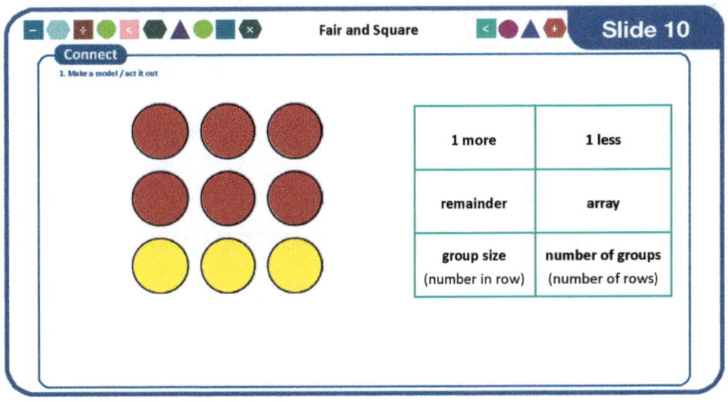

Display slide 10.

The easiest way for children to understand what is happening is to build the arrays. They could be built using double-sided counters or cubes.

- *Choose one pair of calculations. Build the first calculation as an array.*
- *How can you change it to make the second array?*

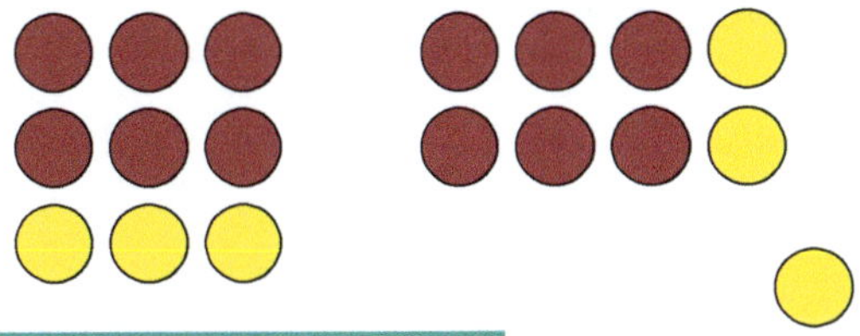

Encourage children to use a different colour for the row they plan to move. This makes it much easier for them to observe how the arrays have changed.

Encourage children to use words from the word bank to explain what has happened.

1 more	**1 less**	**remainder**
array	**group size** (number in a row)	**number of groups** (number of rows)

Connect

Connect

2. Make a table / spot a pattern

Slide 11

Display slide 11.

- *How could you use a table to record what you find?*
- *Could you start with 1 × 1?*
- *Could you go up to 10 × 10?*

Encourage children to draw and complete their own tables.

- *What patterns can you see?*

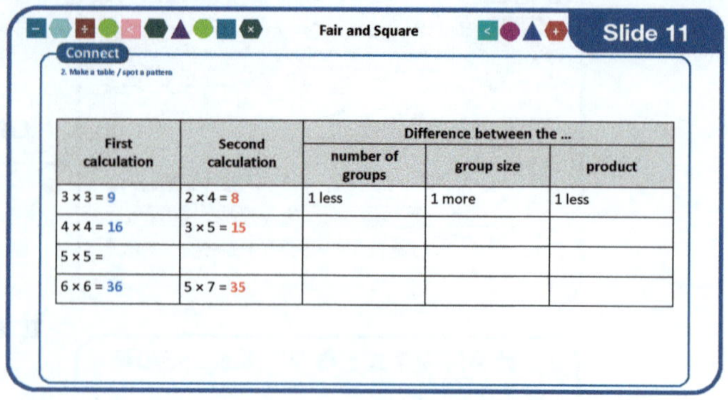

First calculation	Second calculation	Difference between the ...		
		number of groups	group size	product
3 × 3 = 9	2 × 4 = 8	1 less	1 more	1 less
4 × 4 = 16	3 × 5 = 15			
5 × 5 =				
6 × 6 = 36	5 × 7 = 35			

3. Work backwards

Additional resources required: counters or cubes in two different colours

Slide 12

Display slide 12.

- *What arrays could you make using 24 counters or cubes?*

Children could use counters or cubes to make arrays of 1 by 24, 2 by 12, 3 by 8 and 4 by 6.

- *Do any of your arrays fit the pattern you have seen so far?*
- We need an array where the number of rows is 1 less than 5, and the number of columns is 1 more than 5. So the only array that could work is 4 by 6.
- *Can you rearrange your array to match the first calculation?*
- Yes, but I need to add 1 more counter.
- *What arrays would have a product of 48?*

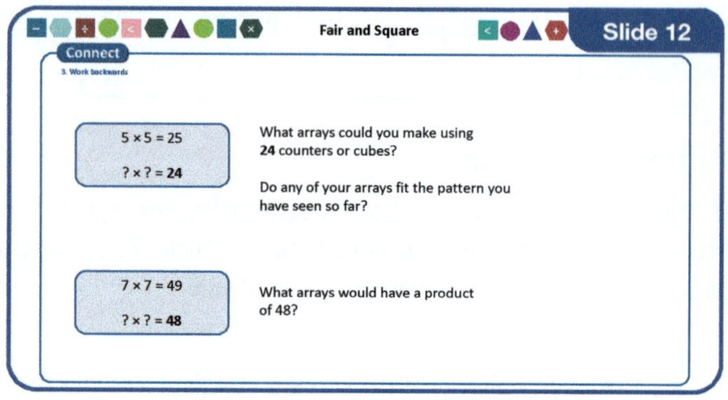

Children might list the factor pairs for 48 or make models.

- The array that fits the pattern is 6 by 8, because 6 is 1 less than 7, and 8 is 1 more than 7.

Collaborate

Refer to the **Collaborate** task of the **Pupil Resource**.

Ask children to form pairs or small groups with others who have used the same strategy. Suggest that they compare their results and ways of working.

- *What was the same about how you worked?*
- *What was different?*

Refer to the **Communicate** task of the **Pupil Resource**.

Slide 13

Display slide 13.

This task provides an opportunity for children to make a poster to advertise their chosen strategy.

Encourage children to use the word bank as they create their posters.

amazing	**strategy**
efficient	**You will be …**
unbelievable	**You'll never need to … again**
accurate	**Just think how …**

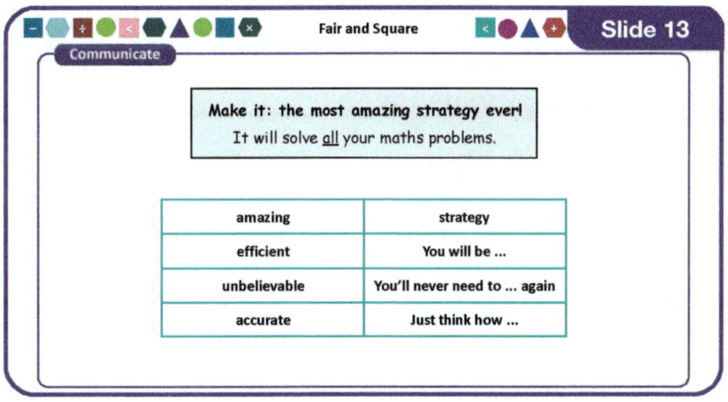

Looking at one another's posters, children can then compare and contrast the different methods of working that they have used.

- *Which strategy do you think is the clearest? Why is that?*

Refer to the **Create** task of the **Pupil Resource**.

Slide 14

Display slide 14.

- *Were there any questions asked at the start of the task that you now want to investigate?*
- *Is there anything new that you are wondering now?*

Children can use the word bank to prompt further questions of their own.

2 more	**2 less**
two-digit numbers	**three numbers**

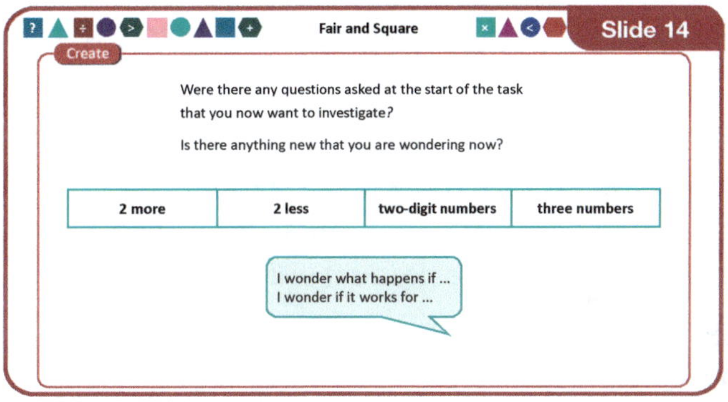

- I wonder what happens if you change the factors by making one 2 less and the other 2 more.
- I wonder if it works with only … numbers.
- I wonder if it works with two-digit numbers.
- I wonder if it works for three numbers multiplied together.

Refer to the **Consider** task of the **Pupil Resource**.

Slide 15

Display slide 15.

Encourage children to think about the strategies they have used. The prompts below may help to get them started.

Cognitive learning domain

- By exploring this pattern, I learned …
- I liked the … strategy because …
- Next time, I will …

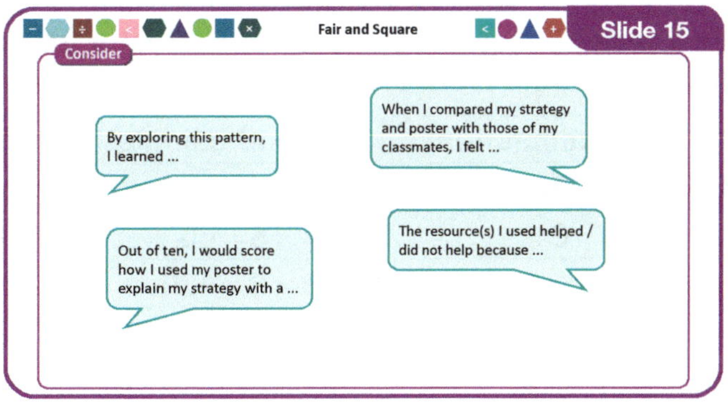

Affective domain

- When I compared my strategy and poster with those of my classmates, I felt ...
- Out of ten, I would score how I used my poster to explain my strategy with a ...
- My ‘aha’ moment was …

Metacognitive domain

- The resource(s) I used helped / did not help because ...
- I found … helpful.
- I found … unhelpful.
- Next time, I will do … differently.

Answers

The expectation is not that children will produce a numerical solution. The objective is for them to use the problem-solving heuristics to explore and explain why the pattern occurs. The children's use of counters or cubes in two colours to show the redistribution provides an opportunity for you to assess their understanding.

Here is an algebraic explanation of why the calculations in each pair have products that differ by 1:

n^2 (square number)

$(n - 1)(n + 1) = n^2 + n - n - 1 = n^2 - 1$

Using two-digit numbers

The pattern is still present when the factors are two-digit numbers. For example:

$100 \times 100 = 10\,000$

$99 \times 101 = 9999$

When the factors are 2 less / 2 more

Some children may assume that, if one factor is 1 less and the other is 1 more, the product will be 2 less. However, the product will actually be 4 less.

For example:

$5 \times 5 = 25$

$3 \times 7 = 21$

Here is an algebraic explanation of why the calculations in each pair have products that differ by 2:

n^2 (square number)

$(n - 2)(n + 2) = n^2 + 2n - 2n - 4 = n^2 - 4$

This is the basis for showing the difference between two squares. If n is the factor used in the first calculation, and x is the amount by which the factors are changed:

$(n - x)(n + x) = n^2 - nx + nx - x^2 = n^2 - x^2$

Multiplying three numbers together (cube numbers)

Children could explore and find a similar pattern for cube numbers. For example, if they make the first factor 1 less, keep the second factor the same, and make the third factor 1 more, the difference in the products will be the same as the number that is being cubed. For example:

First calculation	Second calculation	Difference in product
$3 \times 3 \times 3 = 27$	$2 \times 3 \times 4 = 24$	3 less
$4 \times 4 \times 4 = 64$	$3 \times 4 \times 5 = 60$	4 less
$5 \times 5 \times 5 \times = 125$	$4 \times 5 \times 6 = 120$	5 less
$10 \times 10 \times 10 = 1000$	$9 \times 10 \times 11 = 990$	10 less

Beat the Calculator

Year 4

Introduction

Beat the Calculator provides children with the excitement of a performable trick, made possible by the application of their skills of mental and written multiplication.

Children will seek patterns that are easier to identify if placed in systematically organised tables.

They will then communicate their findings by creating comic strips in which the characters explain how the trick works. This will allow them to combine imagination and accurate mathematical reasoning to convey what they have learned in a simplified and minimal way. It will also lead to observation that using accurate mathematical vocabulary makes it easier to describe patterns effectively.

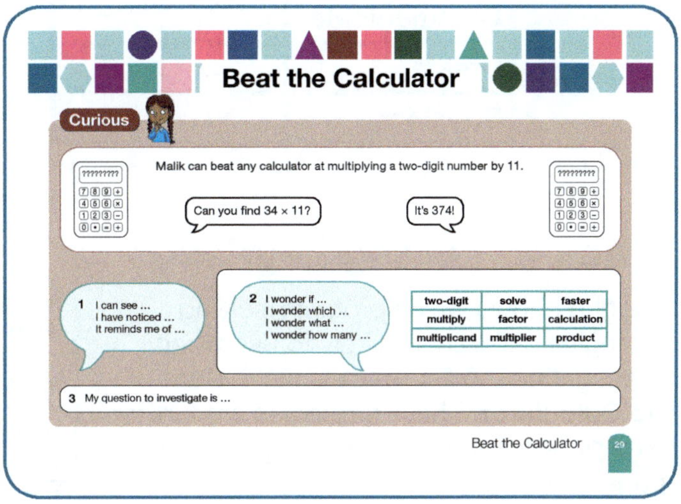

The task opens the door for a wide range of further explorations and investigations, which could lead to the discovery of further mathematical 'tricks'.

This task promotes greater-depth learning in a range of ways, supporting children to identify and articulate underlying mathematical structures. For example, in the **Create** stage, greater depth is promoted by asking children to form their own mathematically accurate questions that are creative and inventive.

This **Teacher Guide** is supported by photocopiable **Pupil Resources** and **Teacher's Presentation** PowerPoint slides for whole-class or group display.

National Curriculum

Beat the Calculator covers these Year 4 National Curriculum objectives.

Multiplication and division

- Recall multiplication and division facts for multiplication tables up to 12×12.
- Solve problems involving multiplying and adding, including using the distributive law to multiply two-digit numbers by one digit, integer scaling problems and harder correspondence problems such as *n* objects are connected to *m* objects.

The 6C cycle

The grid below shows the relative weighting given to each of the 6 Cs in **Beat the Calculator**.

Curious	★★
Connect	★★
Collaborate	★★
Communicate	★★★
Create	★★★
Consider	★★

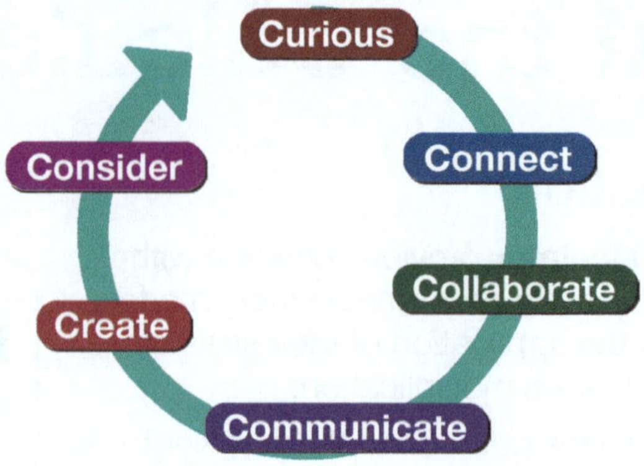

Problem-solving heuristics

The grid shows which problem-solving strategies children are likely to use when investigating the **Connect** tasks in **Beat the Calculator**. These strategies are further exemplified in the **Connect** stage below (pages 93–94).

Draw a diagram	
Make a table / spot a pattern	✓
Make a model / act it out	
Try then improve	
Create a systematic list	
Solve a simpler but related problem	
Work backwards	✓

Key mathematical ideas

Multiplication by 11 of numbers up to 9 is simple and follows a very predictable pattern, making it a favourite times table with many children. However, it takes a little practice to become proficient with multiplying multi-digit numbers by 11.

After an initial demonstration of the 'trick', children recap solving calculations using the distributive law. They then move on to learn how to calculate products for the 11 times table at speed.

To multiply a two-digit number by 11, children will write the two digits with a gap between them. In the gap, they will place the digit sum. If the digit sum is greater than 9, they need to regroup into the hundreds column.

Example 1	Example 2
$15 \times 11 =$ Digit sum: $1 + 5 =$ **6** Place the 6 between the 1 and the 5: $15 \times 11 = 165$	$34 \times 11 =$ Digit sum: $3 + 4 =$ **7** Place the 7 between the 3 and the 4: $34 \times 11 = 374$
Example 3	**Example 4**
$39 \times 11 =$ Digit sum: $3 + 9 =$ **12** Place the 12 between the 3 and the 9, but, rather than making 3129, carry the 1 and add it to the 3: 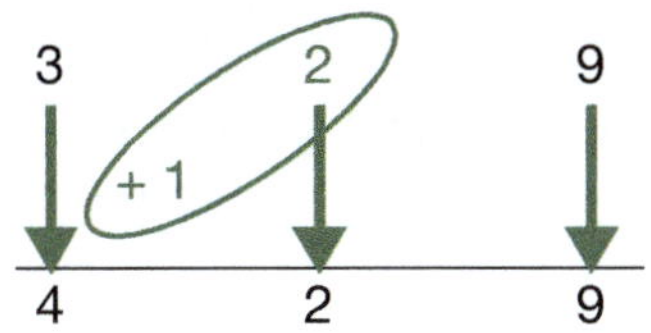 $39 \times 11 = 429$	$67 \times 11 =$ Digit sum: $6 + 7 =$ **13** Place the 13 between the 6 and the 7, but, rather than making 6137, carry the 1 and add it to the 6: 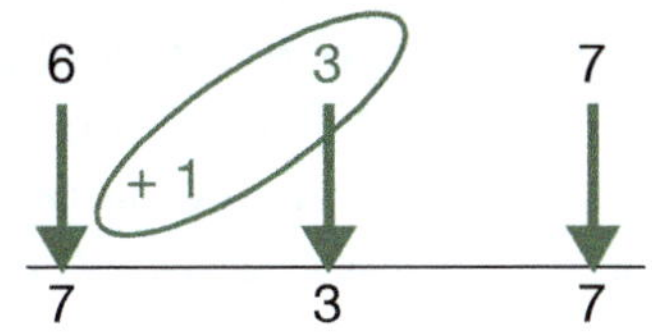 $67 \times 11 = 737$

If the same calculations are solved using long multiplication, it becomes clear why this happens:

$$\begin{array}{r} 15 \\ \times\ 11 \\ \hline 15 \\ 150 \\ \hline 165 \end{array} \qquad \begin{array}{r} 34 \\ \times\ 11 \\ \hline 34 \\ 340 \\ \hline 374 \end{array} \qquad \begin{array}{r} 39 \\ \times\ 11 \\ \hline 39 \\ 390 \\ \hline 429 \\ {\scriptstyle 1\ \ \ \ \ } \end{array} \qquad \begin{array}{r} 67 \\ \times\ 11 \\ \hline 67 \\ 670 \\ \hline 737 \\ {\scriptstyle 1\ \ \ \ \ } \end{array}$$

There is no expectation for children to solve and prove problems using algebra within this task. However, algebra can be used to show the generalised rule and might be interesting for some children.

$ab \times 11 =$

If $a + b < 10$, $ab \times 11 =$

H	T	O
a	$(a + b)$	b

If $a + b > 9$, $ab \times 11 =$

H	T	O
$a + 1$	$(a + b) - 10$	b

Answers to the tasks are given on page 97.

Teaching guidance

If you are using the **Express Pathway**, please turn to pages 93–96.

Teaching guidance for the **Full 6C Pathway** is given below.

The **Teacher's Presentation** PowerPoint file has slides specifically mentioned in this teaching guidance. Some of the slides include notes (within the PowerPoint file) providing 'stumble support' suggestions, that may help you to support children in their problem-solving thinking and enable access for children who are less confident.

Awe and wonder

Additional resources required: calculators

Slide 1

Display slide 1.

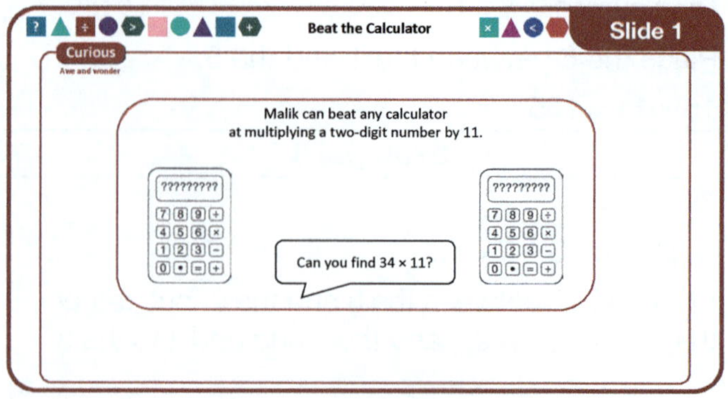

Discuss the statement about Malik, and ask children if they believe he can beat any calculator at multiplying a two-digit number by 11.

Ask children to suggest two-digit numbers you can use to test yourself. Display the calculations, but do not yet show the answers: this may expose the pattern earlier than intended. Provide children with calculators and invite them to race you in finding the answer to each calculation in turn, giving the answers orally. (The time taken for use of a calculator should include entering the equation, not simply the time it takes for the answer to appear after '=' is tapped.)

Provide pairs of children with calculators against which they can race.

- I have noticed that using a calculator isn't always fast.
- I can see that [teacher's name] is good at multiplying by 11.
- I have noticed that if the digits are larger, it takes slightly longer.

Develop and deepen

Continue to solve multiplications, but now display both the calculation and product.

Slide 2

Display slide 2.

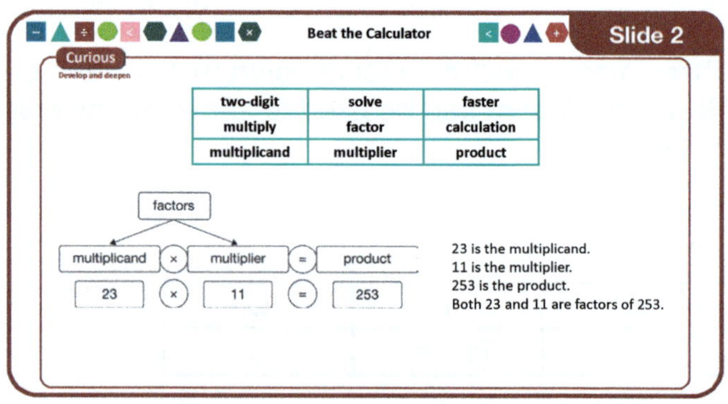

Encourage children to use words from the word bank to develop and deepen their thoughts and questions.

two-digit	**solve**	**faster**
multiply	**factor**	**calculation**
multiplicand	**multiplier**	**product**

The language associated with multiplication may be unfamiliar and can be difficult for children to integrate, as it is not used in everyday life. Discuss the terms 'multiplicand', 'multiplier', 'product' and 'factor'. Use these terms to discuss the calculations you are solving.

Possible lines of enquiry:

- I wonder how it works.

Possible lines of enquiry linked to the word bank:

- I wonder how I can solve these calculations as quickly as Malik.

Possible lines of enquiry that children may raise now, but should be held back until the **Create** task:

- I wonder if you can do the same thing with three- or four-digit numbers.
- I wonder if you can do the same thing with numbers that have zeros as digits.

Consider whether / how to collate and display all the suggested questions, and whether / how to steer children towards selecting an appropriately challenging question.

Pupil Resources for Full and Express Pathways

Distribute the **Pupil Resources** for **Beat the Calculator** to the children to help support them in their exploration of the different tasks. Use either of the **Curious Pupil Resources** depending on whether you are following the **Full Pathway** or the **Express Pathway**. The **Pupil Resources** for the remaining 5Cs are common to both pathways.

To provide additional flexibility in the classroom, each of the **Pupil Resources** is also provided in the **Teacher's Presentation** PowerPoint file (see slides 3–6), so you can display these at any time if you wish.

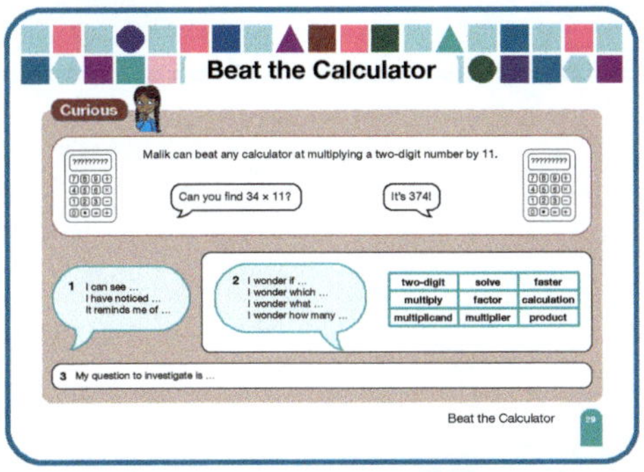

or

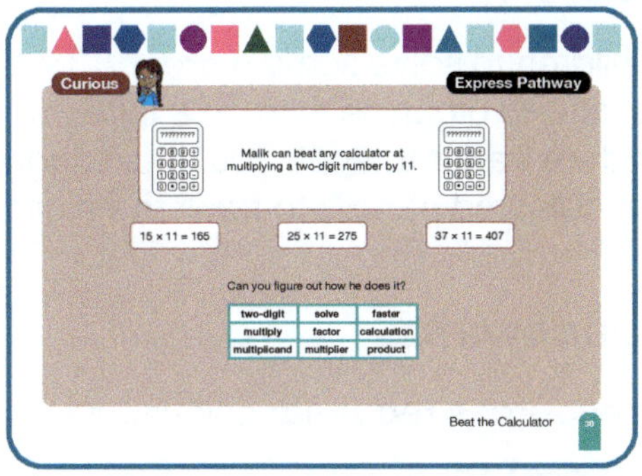

then

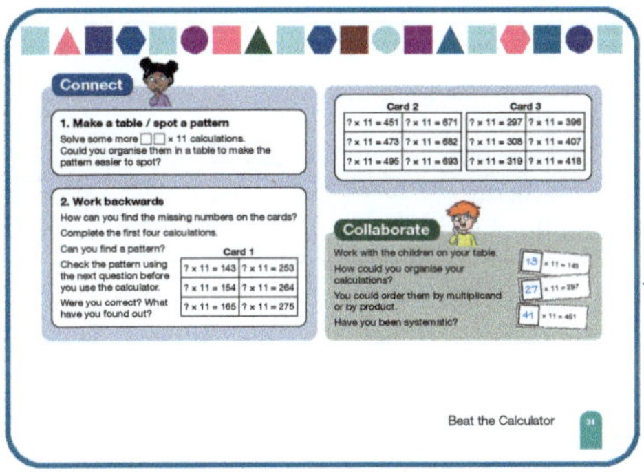

and

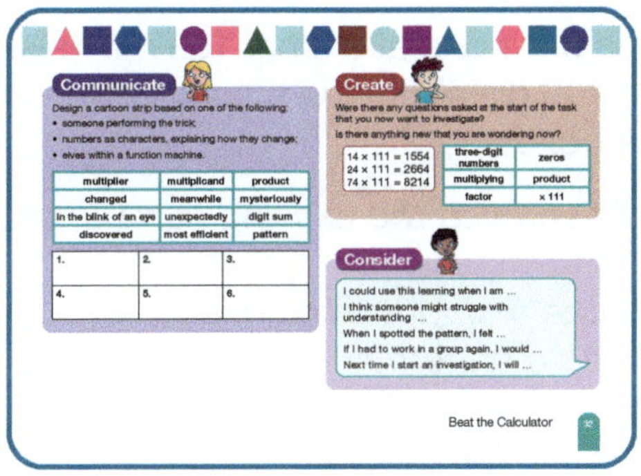

Express Pathway

If you are following the **Express Pathway**, spend a few minutes discussing the **Curious Pupil Resource** with the children. (This is slide 4, so you could display it if you wish.) Use some of the prompts on the previous page to help.

Two tasks involving different strategies are suggested for children to make connections.

1. Make a table / spot a pattern

Additional resources required: calculators, coloured pencils

Slide 7

Display slide 7.

Ask children to find the products of the multiplications using the distributive law.

For example:

$$13 \times 11 = (10 \times 11) + (3 \times 11) = 110 + 33 = 143$$

$$29 \times 11 = (20 \times 11) + (9 \times 11) = 220 + 99 = 319$$

Then ask them to check their answers using calculators.

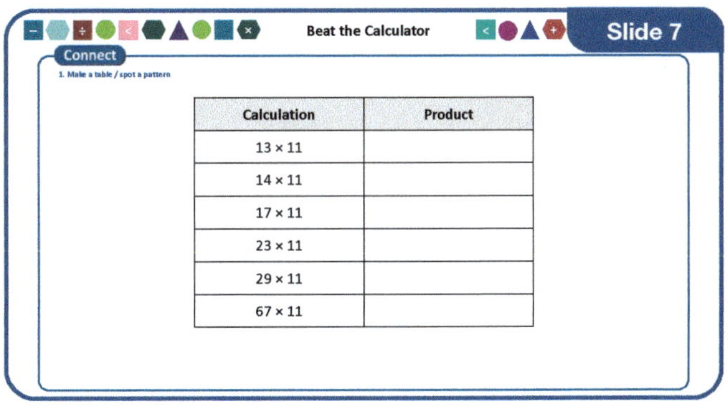

Calculation	Product
13 × 11	
14 × 11	
17 × 11	
23 × 11	
29 × 11	
67 × 11	

⚠ Ensure that children are confident with using a calculator.

Slide 8

Display slide 8. This reveals the correct products, with colour coding that indicates the pattern children should have spotted.

- *What is the pattern?*
- *Where does the pattern change?*
- *What is different about the digits involved in that calculation?*

Challenge children to solve some more □□ × 11 calculations, and to organise them in a table to make the pattern easier to spot. You could suggest that they use coloured pencils to track the digits.

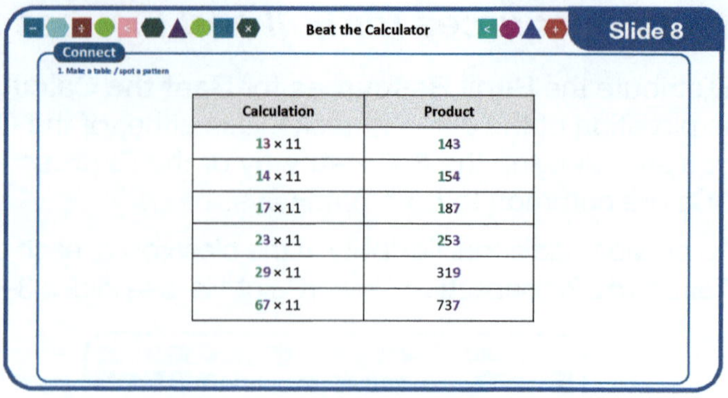

2. Work backwards

Additional resources required: calculators

Slides 9, 10 and 11

Display slides 9–11 one at a time. You may also wish to print and distribute them so children can fill in the boxes.

- *How could you find the missing numbers on the cards?*
- *Complete the first four calculations.*
- *Can you find a pattern?*
- *Check the pattern using the next question before you use the calculator.*
- *Were you correct? What have you found out?*

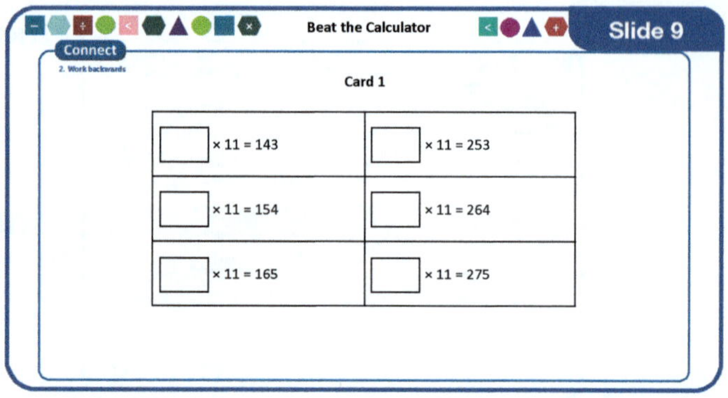

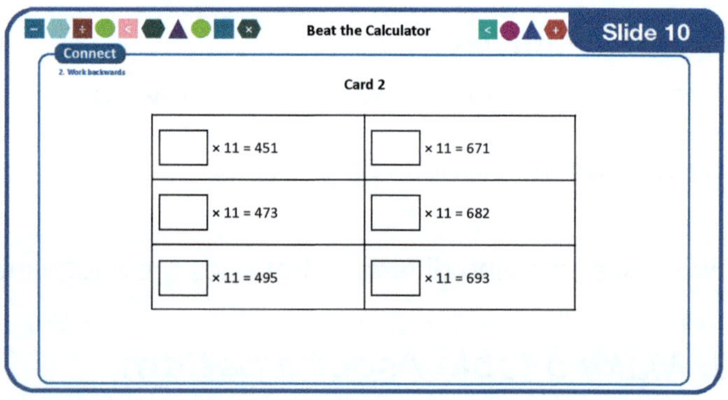

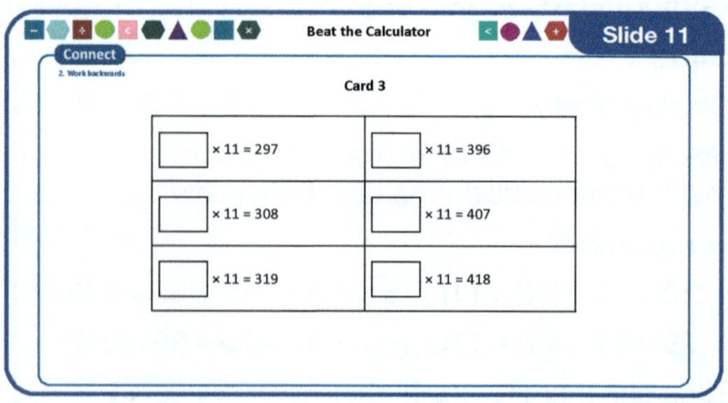

Refer to the **Collaborate** task of the **Pupil Resource**.

Children's access to more completed calculations will give them more information to construct into tables that show the pattern in the 11 times table.

If you have printed and distributed slides 9–11, ask children to cut up the tables to create a slip of paper for each accurately completed calculation. If not, children could copy them onto paper strips. Ask children to work with others on their table to organise the calculations logically.

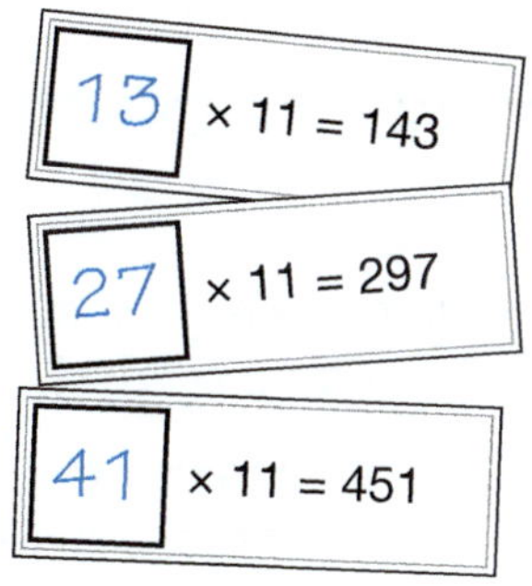

- *How could you organise your calculations?*
- *Could you use a table?*
- *Could you order them by multiplicand or by product?*
- *Have you been systematic?*

Refer to the **Communicate** task of the **Pupil Resource**.

Slide 12

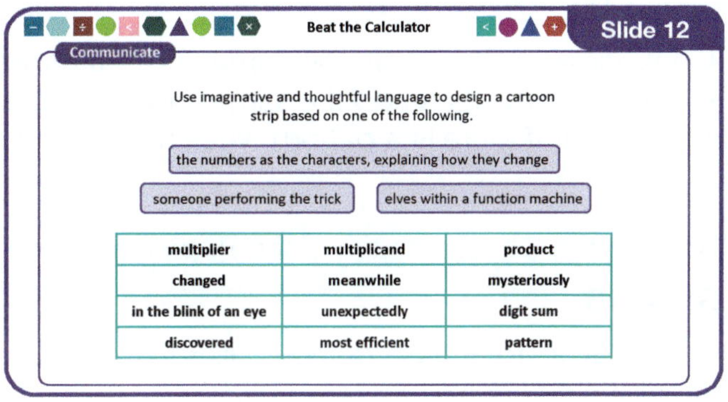

Display slide 12. This task provides an opportunity for the children to design a cartoon strip that uses imaginative and thoughtful language to show how the 'trick' works.

They could base their cartoons on one of the following premises:

- someone performing the trick;
- the numbers as the characters, explaining how they change when multiplied by 11;
- elves within a function machine.

Encourage them to use words from the word bank when they write.

multiplier	**multiplicand**	**product**
changed	**meanwhile**	**mysteriously**
in the blink of an eye	**unexpectedly**	**digit sum**
discovered	**most efficient**	**pattern**

Slide 13

Print and distribute slide 13 (a panel structure on which children can draw their comic strips).

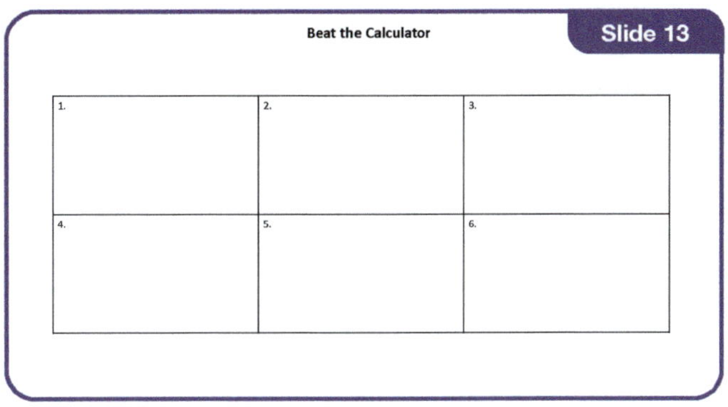

Create

Refer to the **Create** task of the **Pupil Resource**.

Slide 14

Display slide 14.

- *Were there any questions asked at the start of the task that you now want to investigate?*
- *Is there anything new that you are wondering now?*

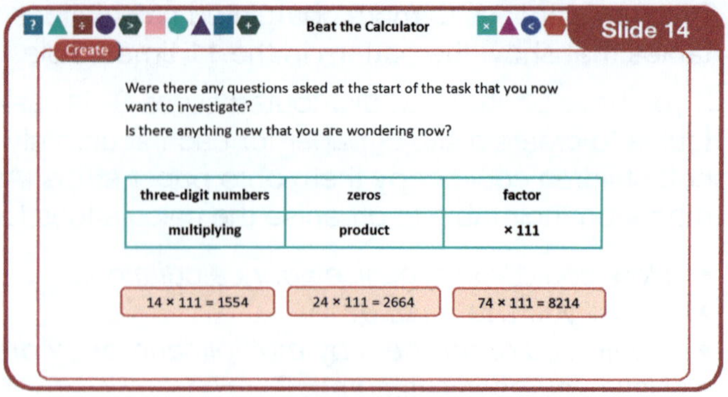

Children can use the word bank to prompt further questions of their own.

three-digit numbers	**zeros**
factor	**multiplying**
product	**× 111**

- I wonder if you can do the same thing with three- or four-digit numbers.
- I wonder if you can do the same thing with numbers that have zeros as digits.
- I wonder if there is a pattern when multiplying by 111.

Consider

Refer to the **Consider** task of the **Pupil Resource**.

Slide 15

Display slide 15 and encourage children to think about the strategies that they have used. The prompts below may help to get them started.

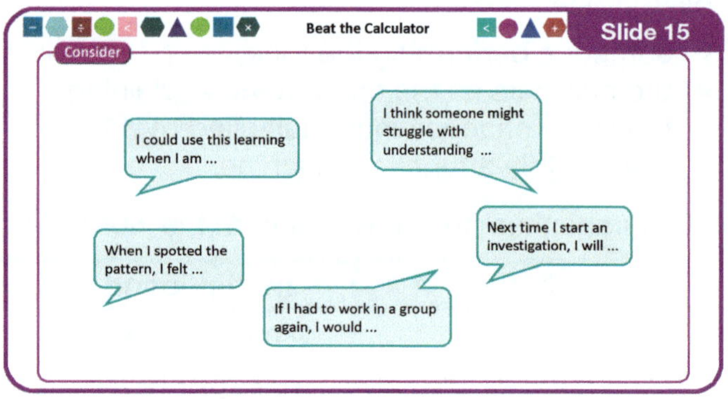

Cognitive learning domain

- I could use this learning when I am ...
- I think someone might struggle with understanding ...
- I noticed that ...

Affective domain

- I really enjoyed ...
- When I spotted the pattern, I felt ...
- Working in a group made me feel ...

Metacognitive domain

- If I had to work in a group again, I would ...
- Next time I start an investigation, I will ...
- If I had to do this task again, I would ...
- I organised my work by ...

Answers

Multiplying a two-digit number by 11

First, find the digit sum of the two-digit number.

If the digit sum is less than 10, the product can be written as: [first digit][digit sum][second digit].

Example: 27×11 Digit sum: $2 + 7 = 9$ $27 \times 11 = 297$

If the digit sum is greater than 9, the product can be written as [first digit + 1][second digit of digit sum][second digit].

Example: 59×11 Digit sum: $5 + 9 = 14$ $59 \times 11 = 649$

These rules work even if the multiplicand has a zero in the ones place.

Multiplying a three-digit number by 11

This involves finding the digit sum of each adjacent pair of digits in the multiplicand.

If the digit sums are all less than 10, the product can be written as: [first digit][sum of first and second digits][sum of second and third digits][third digit].

Example: 234×11 Digit sum of first and second digits: $2 + 3 = 5$

Digit sum of second and third digits: $3 + 4 = 7$ $234 \times 11 = 2574$

If either of the digit sums is greater than 9, the tens need to be carried to the next digit.

Example: 572×11 Digit sum of first and second digits: $5 + 7 = 12$

Digit sum of second and third digits: $7 + 2 = 9$ $572 \times 11 = 6292$

These rules work even if the multiplicand has a zero in the tens or ones place.

Multiplying a four-digit number by 11

The method for this is the same as for three-digit numbers.

Example: 2341×11 Digit sum of first and second digits: $2 + 3 = 5$

Digit sum of second and third digits: $3 + 4 = 7$

Digit sum of third and fourth digits: $4 + 1 = 5$ $2341 \times 11 = 25\,751$

Multiplying a two-digit number by 111

First, find the digit sum of the two-digit number.

If the digit sum is less than 10, the product can be written as: [first digit][digit sum][digit sum][second digit].

Example: 27×111 Digit sum: $2 + 7 = 9$ $27 \times 111 = 2997$

If the digit sum is greater than 9, the product can be written as [first digit + 1][total of the two digits in the digit sum][second digit of digit sum][second digit].

Example: 59×111 Digit sum: $5 + 9 = 14$ $59 \times 111 = 6549$

These rules work even if the multiplicand has a zero in the ones place.

Stabilising Snowflakes

Year 3

Introduction

Stabilising Snowflakes prompts children to explore equivalent fractions in more depth, as they are faced with fractions with different denominators. Although children do not add fractions with different denominators In Lower Key Stage 2, this task encourages them to consider other ways to create three equal pairs of fractions, in the context of balancing the sides of three intersecting lines in a snowflake.

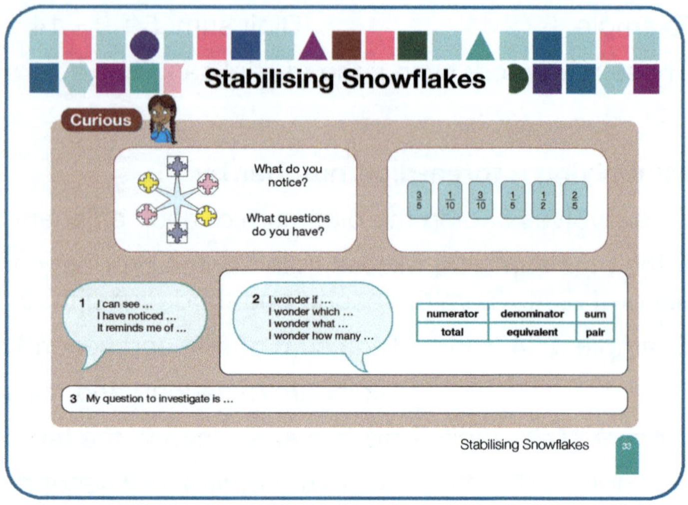

The solution can be found using fraction walls and bar models, allowing children to connect their knowledge of addition structures to fractions. They may also discover patterns in the given fractions, noticing that the majority are divisible by 10 and can be converted easily to tenths.

This task promotes greater-depth learning in a range of ways, supporting children to identify and articulate underlying mathematical structures. For example, in the **Connect** stage, greater depth is promoted by asking children to explore a range of different mathematical resources and models, choose appropriate resources for the task, and explain their choice.

This **Teacher Guide** is supported by photocopiable **Pupil Resources** and **Teacher's Presentation** PowerPoint slides for whole-class or group display.

National Curriculum

Stabilising Snowflakes covers these Year 3 National Curriculum objectives.

Fractions

- Recognise and show, using diagrams, equivalent fractions with small denominators.
- Add and subtract fractions with the same denominator within one whole [for example, $\frac{5}{7} + \frac{1}{7} = \frac{6}{7}$].

The 6C cycle

The grid on the right shows the relative weighting given to each of the 6 Cs in **Stabilising Snowflakes**.

Curious	★★
Connect	★★★
Collaborate	★
Communicate	★
Create	★★
Consider	★★

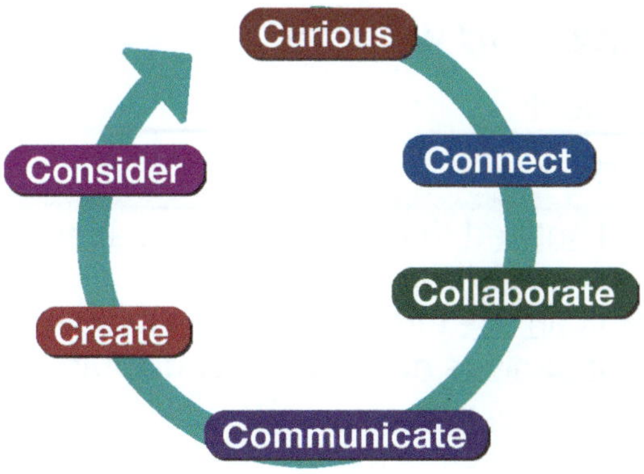

Problem-solving heuristics

The grid shows which problem-solving strategies children are likely to use when investigating the **Connect** tasks in **Stabilising Snowflakes**. These strategies are further exemplified in the **Connect** stage below (pages 102–104).

Draw a diagram	✓
Make a table / spot a pattern	✓
Make a model / act it out	✓
Try then improve	
Create a systematic list	
Solve a simpler but related problem	
Work backwards	

Key mathematical ideas

Generally (if following the National Curriculum for England), children in Lower Key Stage 2 are not taught how to add fractions with different denominators. In this task, they will discover other ways of balancing the fraction pairs to ensure the two fractions in each of the three pairs are equal in value.

These methods include using concrete manipulatives and 2D representations that they relate to the numerals, and using known number bonds about the factors of ten to convert denominators, which allows children to compare fractions directly. Children select their preferred strategy and then compare it to a different one by communicating with their peers.

Answers to the tasks are given on page 106.

Teaching guidance

If you are using the **Express Pathway**, please turn to pages 101–106.

Teaching guidance for the **Full 6C Pathway** is given below.

The **Teacher's Presentation** PowerPoint file has slides specifically mentioned in this teaching guidance. Some of the slides include notes (within the PowerPoint file) providing 'stumble support' suggestions, that may help you to support children in their problem-solving thinking and enable access for children who are less confident.

Awe and wonder

Slide 1

Display slide 1.

- I noticed that there are six shapes around the outside.
- I can see three pairs of coloured shapes.
- It reminds me of a spider diagram.

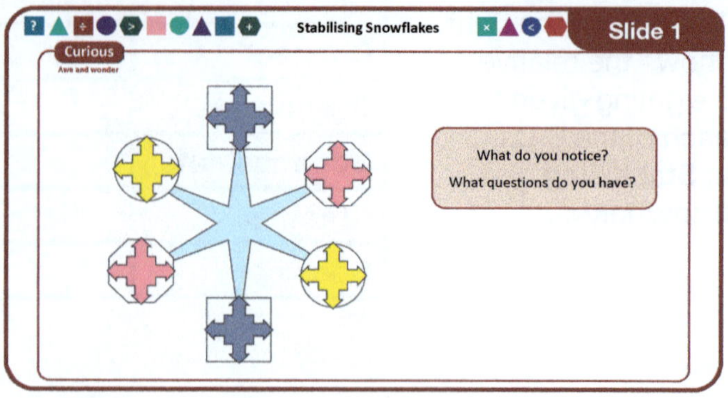

Slide 2

Display slide 2.

- I can see that there are six fractions.
- I have noticed that the fractions do not have the same numerator / denominator.
- I can see that there are two unit fractions.

⚠ Watch out for children who are not confident in the use of 'numerator' and 'denominator'. If needed, recap the key vocabulary for fractions.

Develop and deepen

Slide 3

Display slide 3.

Discuss the central problem of the task: children have to find a way to pair the fractions so the sums of the three pairs are the same.

Encourage children to use words from the word bank to develop and deepen their thoughts and questions.

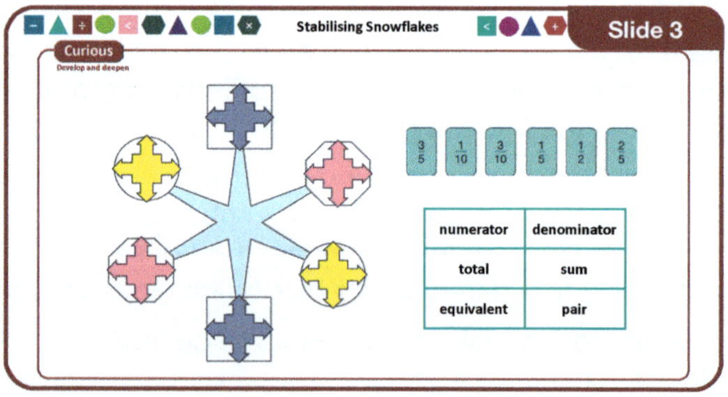

numerator	**denominator**	**sum**
total	**equivalent**	**pair**

Possible lines of enquiry:

- I wonder if I can make each pair add up to the same value as the others.

Possible lines of enquiry linked to the word bank:

- I wonder what would the total of all the fractions be.
- I wonder what the value of the greatest pair would be.
- I wonder how I can add fractions with different denominators.

Possible lines of enquiry that children may raise now, but should be held back until the **Create** task:

- I wonder whether I can find other fraction pairs with the same totals.
- I wonder if I could create my own puzzle like this.

Consider whether / how to collate and display all the suggested questions, and whether / how to steer children towards selecting an appropriately challenging question.

Pupil Resources for Full and Express Pathways

Distribute the **Pupil Resources** for **Stabilising Snowflakes** to the children to help support them in their exploration of the different tasks. Use either of the **Curious Pupil Resources** depending on whether you are following the **Full Pathway** or the **Express Pathway**. The **Pupil Resources** for the remaining 5Cs are common to both pathways.

To provide additional flexibility in the classroom, each of the **Pupil Resources** is also provided in the **Teacher's Presentation** PowerPoint file (see slides 4–7), so you can display these at any time if you wish.

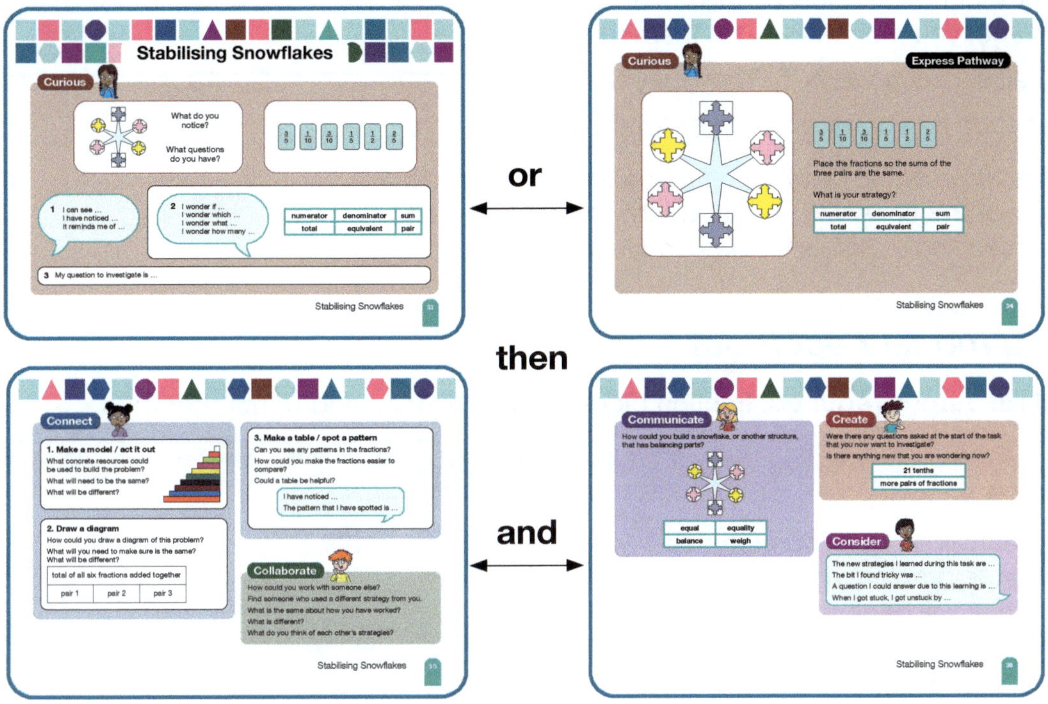

Express Pathway

If you are following the **Express Pathway**, spend a few minutes discussing the **Curious Pupil Resource** with the children. (This is slide 5, so you could display it if you wish.) Use some of the prompts on the previous page to help.

Three tasks involving different strategies are suggested for children to make connections.

Slide 8

Print and distribute copies of slide 8 (a fraction wall). Children may find this a useful reference for any of the **Connect** tasks.

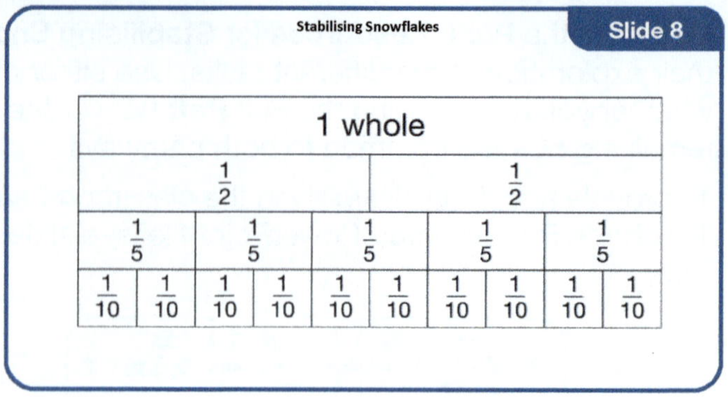

Note: children should not be expected to find the totals of the fractions in their current forms (because they have different denominators). If they have not found the totals by the end of the **Connect** stage, you may wish to tell them that the sum of all six fractions is $\frac{21}{10}$. They may then reason that each pair of fractions must total $\frac{7}{10}$.

- *There are three equal pairs. What is the total of each pair?*
- Each pair of fractions will be equivalent to seven-tenths.
- *Now that we know this, what else do we know?*

1. Make a model / act it out

Additional resources required: fraction blocks, fraction strips and/or number rods

Slide 9

Display slide 9.

Some children will be most comfortable approaching the problem using concrete resources. They will be able to manipulate the fraction pieces to find each of the fractions involved in the problem.

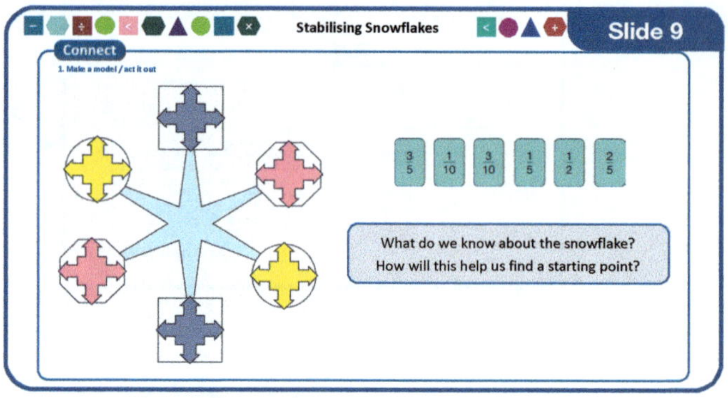

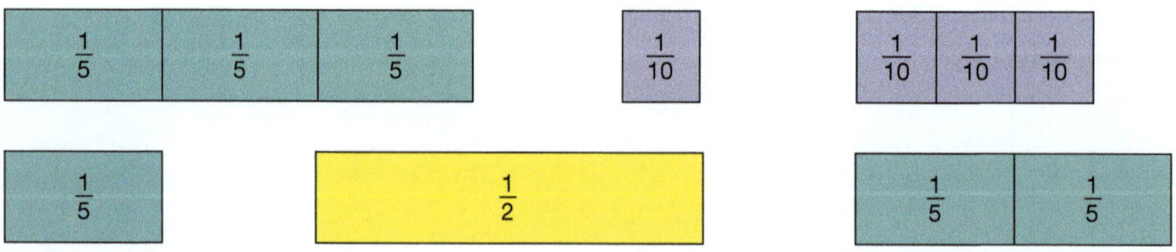

The six fractions, represented using fraction strips

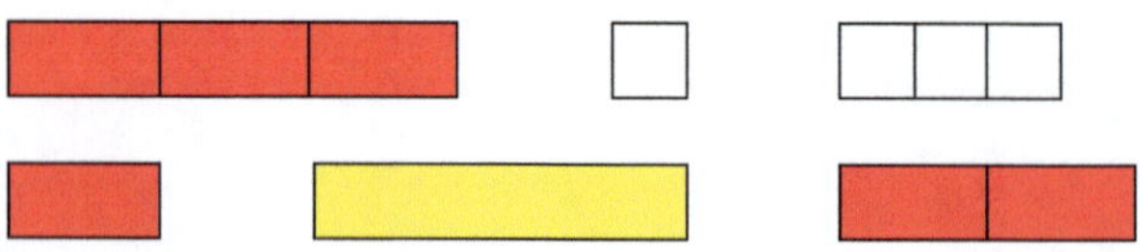

The six fractions, represented using number rods (1-rods for tenths, 2-rods for fifths, 5-rods for halves)

⚠ Ensure children are using fraction pieces that are proportional to each other.

Some children may choose to use the fraction wall to convert all the pieces to tenths before working with them, for example, converting one-half into five-tenths.

Children can then find three pairs of fractions that are of equal size (and therefore of equal value).

Slide 10

Display slide 10, which reveals two of the three equal pairs. Children should now be able to find the third.

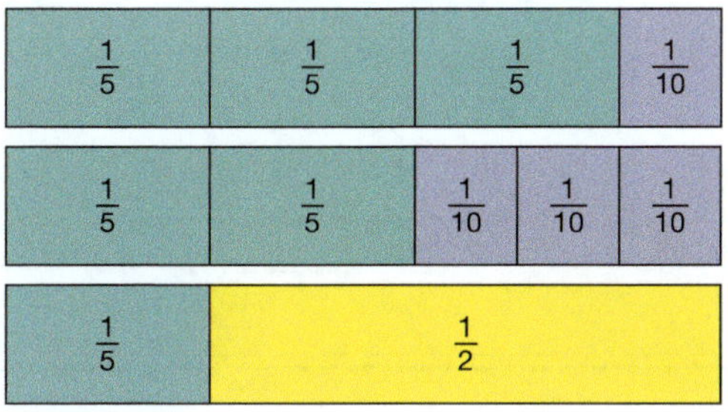

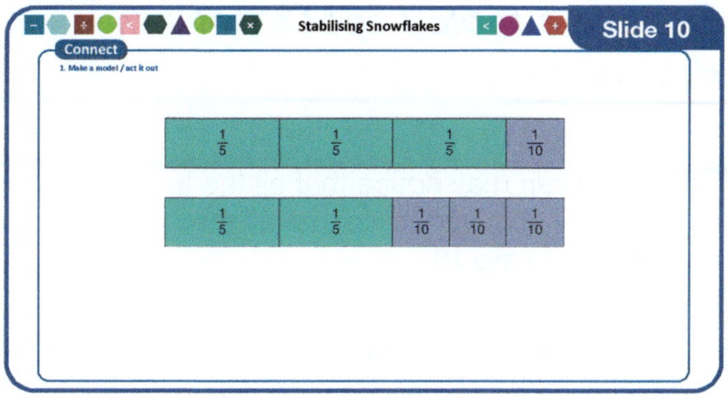

- *What three equal pairs of fractions have you found?* ($\frac{3}{5} + \frac{1}{10}$, $\frac{2}{5} + \frac{3}{10}$ and $\frac{1}{5} + \frac{1}{2}$.)

2. Draw a diagram

Additional resources required: squared paper

Some children may choose to draw 2D representations (for example, bar models) to solve the problem.

Children have been told that the three fraction pairs are equal.

Total of all six fractions added together		
Pair 1	Pair 2	Pair 3

Reveal that the total of all of the fractions is 21 tenths.

$\frac{21}{10}$		
Pair 1	Pair 2	Pair 3

Children will be able to find the value of each pair.

$\frac{21}{10}$		
$\frac{7}{10}$	$\frac{7}{10}$	$\frac{7}{10}$

They can then use the fraction wall to find which combinations of the fractions are equivalent to seven-tenths.

$\frac{3}{5} + \frac{1}{10}$ $\frac{2}{5} + \frac{3}{10}$ $\frac{1}{5} + \frac{1}{2}$

Connect

3. Make a table / spot a pattern

Slide 11

Display slide 11.

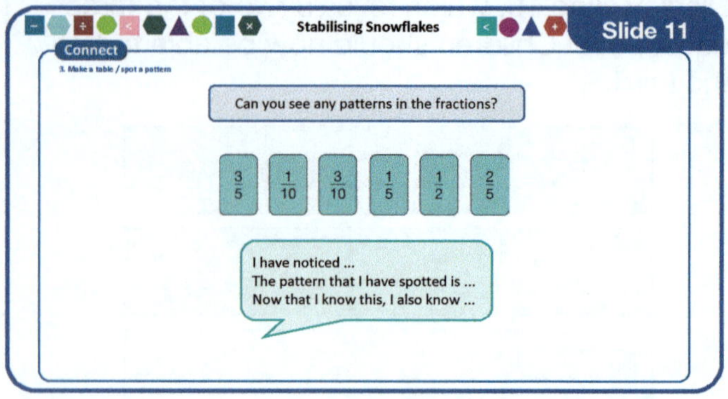

Some children may notice that all the fractions can be easily converted so that all the denominators are 10.

$\frac{3}{5} = \frac{6}{10}$ $\frac{1}{10}$ $\frac{3}{10}$

$\frac{1}{5} = \frac{2}{10}$ $\frac{1}{2} = \frac{5}{10}$ $\frac{2}{5} = \frac{4}{10}$

⚠ Watch out for children who do not recognise that the different fractions need to be multiplied by different numbers to make all the denominators 10 (that is, some by 2 and some by 5).

Children may find it helpful to record the fractions in a table. Encourage them to think about using a systematic order.

Original fraction	Converted to tenths	I multiplied the numerator and denominator by ...
$\frac{1}{10}$	$\frac{1}{10}$	1
$\frac{1}{5}$	$\frac{2}{10}$	2
$\frac{3}{10}$	$\frac{3}{10}$	1
$\frac{2}{5}$	$\frac{4}{10}$	2
$\frac{1}{2}$	$\frac{5}{10}$	5
$\frac{3}{5}$	$\frac{6}{10}$	2

⚠ Once the fractions have been converted, some children may need help to find a starting point: you may need to make links to whole numbers first. For example, ask children how they could find pairs that add up to the same total using the numbers 6, 1, 5, 3, 2 and 4.

Children can then put the fractions into pairs so the sums of the pairs are the same.

Some children may begin to spot number bonds or use a 'try then improve' strategy to find the combinations. Others may reason that the total of the six fractions is 21 tenths, so seek three pairs that each total seven-tenths.

$\frac{6}{10} + \frac{1}{10}$ $\frac{4}{10} + \frac{3}{10}$ $\frac{2}{10} + \frac{5}{10}$

Refer to the **Collaborate** task of the **Pupil Resource**.

There are two main ways for children to collaborate during this task.

At the **Connect** stage, they could partner with a child who has chosen a strategy different from theirs, to compare and contrast their methods and results.

Alternatively, once the task has been solved, they could collaborate in pairs or small groups to complete the **Communicate** stage as part of a Design and Technology project.

Refer to the **Communicate** task of the **Pupil Resource**.

Slide 12

Display slide 12.

equal	equality
balance	weigh

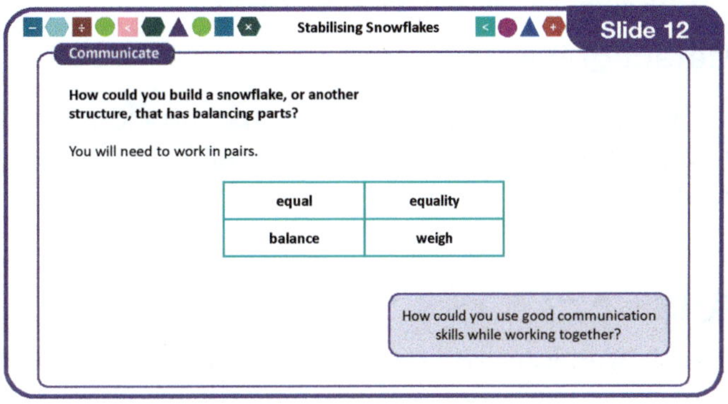

This task provides an opportunity for children to demonstrate the equality of the three pairs of fractions. This could form part of a Design and Technology project in which they build balanced snowflakes. Encourage them to use words from the word bank as they plan and build their snowflakes.

During the making process, children may have the opportunity to meet the following Design and Technology objective from the Key Stage 2 National Curriculum.

- Select from and use a wider range of tools and equipment to perform practical tasks [for example, cutting, shaping, joining and finishing], accurately.

Refer to the **Create** task of the **Pupil Resource**.

Slide 13

Display slide 13.

- *Were there any questions asked at the start of the task that you now want to investigate?*
- *Is there anything new that you are wondering now?*

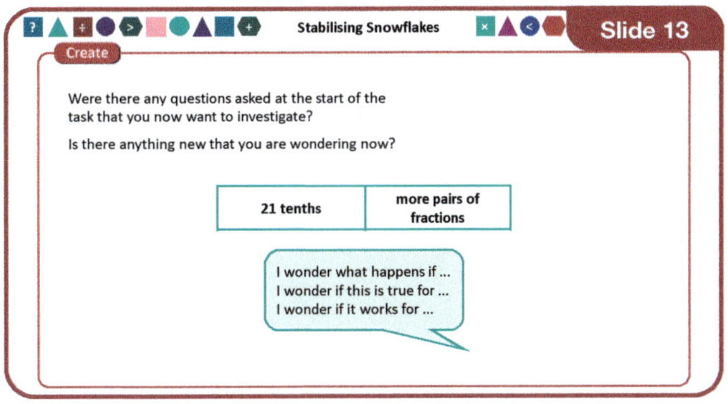

Children can use the word bank to prompt further questions of their own.

21 tenths	more pairs of fractions

- <u>I wonder</u> how else I can make 21 tenths.
- <u>I wonder</u> whether I can find other fraction pairs with the same totals.
- <u>I wonder</u> if I could create my own puzzle like this.
- <u>I wonder</u> if I can have … pairs instead of three.

Consider

Refer to the Consider task of the **Pupil Resource.**

Slide 14

Display slide 14 and encourage children to think about the strategies that they have used. The prompts below may help to get them started.

Cognitive learning domain

- The new strategies I learned during this task are ...
- The bit I found tricky was ...
- A question I could answer due to this learning is ...

Affective domain

- The fun bit of my learning was … because …
- I said 'aha' when … because …
- I was surprised about … because …

Metacognitive domain

- The next time I answer a curious question, the target I would set myself for improvement is
- Out of ten, I would score how well I worked with other children with a ...
- When I got stuck, I got unstuck by ...

Answers

The six fractions add up to 21 tenths. So each pair of fractions must total seven-tenths.

The three pairs of fractions are:

$\frac{3}{5} + \frac{1}{10}$ $\frac{2}{5} + \frac{3}{10}$ $\frac{1}{5} + \frac{1}{2}$

One arrangement of these fractions in the snowflake is:

Queen Bee

Year 3

Year 4

Introduction

Queen Bee gives children an opportunity to explore fractions as equal parts of a whole, and to draw on their understanding of symmetry and properties of shapes.

The challenge is to 'decorate' a beehive, by colouring it in different colours. The hive is made up of 49 identical hexagons. The central hexagon contains the Queen Bee, so this one cannot be decorated. The rules that children must follow are:

- there must be equal numbers of hexagons in each colour;
- no adjacent hexagons can be the same colour.

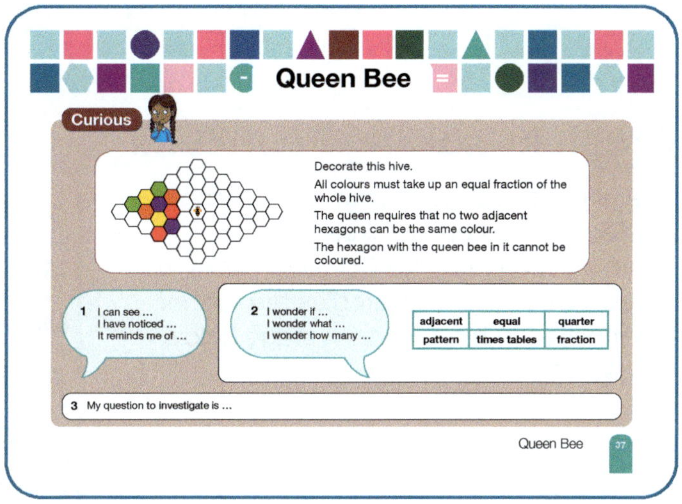

This task promotes greater-depth learning in a range of ways, supporting children to identify and articulate underlying mathematical structures. For example, in the **Consider** stage, greater depth is promoted by asking children to identify learning behaviours and strategies that were and were not useful, and to reflect on what was and was not engaging about the task.

This **Teacher Guide** is supported by photocopiable **Pupil Resources** and **Teacher's Presentation** PowerPoint slides for whole-class or group display.

National Curriculum

Queen Bee covers these Year 3 and Year 4 National Curriculum objectives.

Year 3 Fractions

- Recognise, find and write fractions of a discrete set of objects: unit fractions and non-unit fractions with small denominators.
- Recognise and show, using diagrams, equivalent fractions with small denominators.
- *Continue to recognise fractions in the context of parts of a whole, numbers, measurements, a shape, and unit fractions as a division of a quantity.* [non-statutory guidance]

Year 4 Fractions

- Solve problems involving increasingly harder fractions to calculate quantities, and fractions to divide quantities, including non-unit fractions where the answer is a whole number.
- Recognise and show, using diagrams, families of common equivalent fractions.
- Recognise and write decimal equivalents to $\frac{1}{4}$, $\frac{1}{2}$, $\frac{3}{4}$.

The 6C cycle

The grid to the right shows the relative weighting given to each of the 6 Cs in **Queen Bee**.

Curious	★
Connect	★★
Collaborate	★
Communicate	★★
Create	★★
Consider	★★★

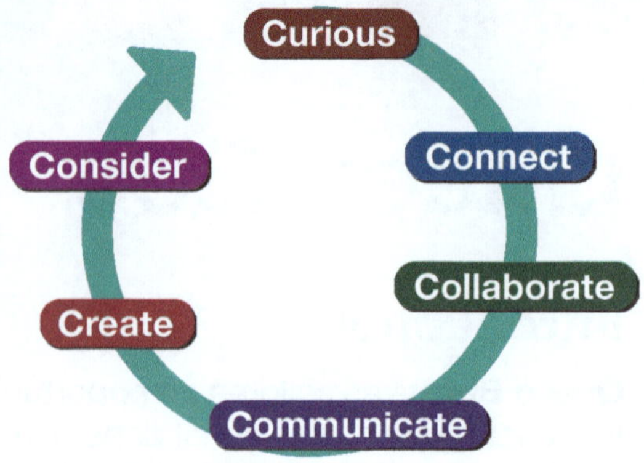

Problem-solving heuristics

The grid shows which problem-solving strategies children are likely to use when investigating the **Connect** tasks in **Queen Bee**. These strategies are further exemplified in the **Connect** stage below (pages 111–113).

Draw a diagram	✓
Make a table / spot a pattern	✓
Make a model / act it out	
Try then improve	✓
Create a systematic list	
Solve a simpler but related problem	
Work backwards	

Key mathematical ideas

Working systematically is key to solving this problem. There is more than one solution, so children can compare both their strategies and their outcomes with their peers.

Once children have established that there are 48 hexagons that must be decorated, they can start to use what they know about the factors of 48 to think about how many colours could be used. They will use known multiplication facts, such as $6 \times 8 = 48$, and derive new related facts, such as $3 \times 16 = 48$. To support systematic working, children could use a table. This will prompt consideration of the order in which different numbers of colours are explored.

When applying the second rule (no adjacent hexagons can be the same colour), an element of 'try then improve' is likely to be needed. Children should identify a suitable starting point and justify their choice. It will become clear that hexagons with six adjacent hexagons provide more information than hexagons with only two.

Answers to the tasks are given on page 116.

Teaching guidance

If you are using the **Express Pathway**, please turn to pages 110–115.

Teaching guidance for the **Full 6C Pathway** is given below.

The **Teacher's Presentation** PowerPoint file has slides specifically mentioned in this teaching guidance. Some of the slides include notes (within the PowerPoint file) providing 'stumble support' suggestions, that may help you to support children in their problem-solving thinking and enable access for children who are less confident.

Awe and wonder

Slide 1

Display slide 1, allowing time for discussion and observations as prompted below.

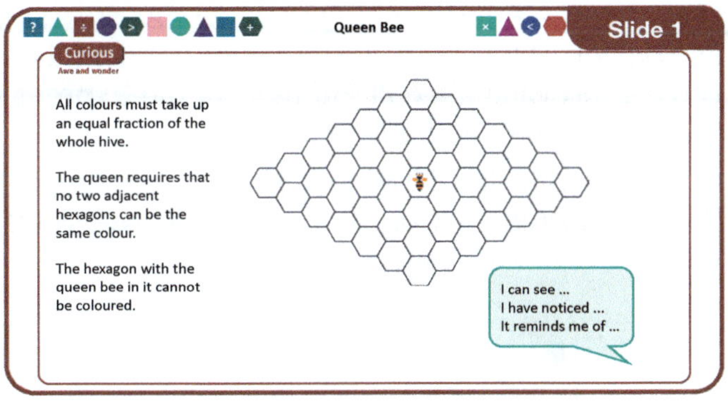

- I can see that the middle cannot be coloured. There is a bee there.
- I have noticed there are 49 hexagons.
- I have noticed that the hexagons are in columns of 1, 2, 3, 4, 5, 6, 7, 6, 5, 4, 3, 2 and 1, from left to right.
- It reminds me of a honeycomb.
- It reminds me of a diamond.
- I have noticed that there are seven hexagons along each side.
- I have noticed that some hexagons are touching only two others but some are touching six others.

Children may use the word 'diamond' rather than 'rhombus'. You may choose to correct their terminology in the interests of their understanding of shape properties.

Develop and deepen

Slide 2

Display slide 2.

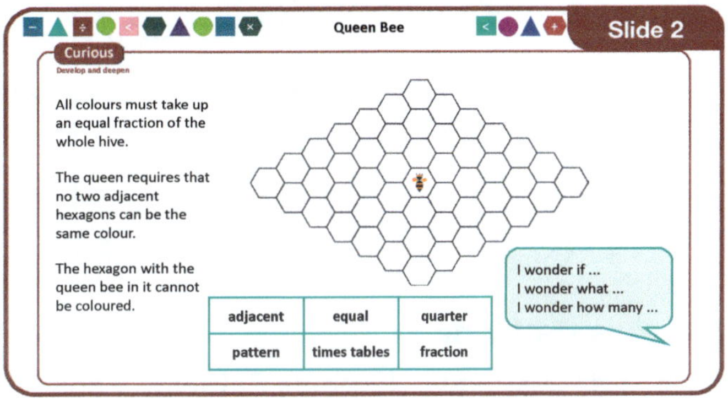

Encourage children to use words from the word bank to develop and deepen their thoughts and questions.

adjacent	**equal**	**quarter**
pattern	**times tables**	**fraction**

Possible lines of enquiry:

- I wonder how many different colours I could use.

Possible lines of enquiry linked to the word bank:

- I wonder what patterns I could create.
- I wonder if I could make a symmetrical pattern in the hive.
- I wonder if I could create a hive with no adjacent hexagons of the same colour.
- I wonder if I could use my times tables facts to help me to decide how many hexagons should be in each colour.
- I wonder if I could create a hive with no adjacent hexagons of the same colour and with equal fractions of each colour.
- I wonder what the largest fraction for a colour would be if adjacent hexagons were different colours and the fractions of the colours used were equal.

Possible lines of enquiry that children may raise now, but should be held back until the **Create** task:

- I wonder what would happen if the hive was built from triangles or squares.

Consider whether / how to collate and display all the suggested questions, and whether / how to steer children towards selecting an appropriately challenging question.

Pupil Resources for Full and Express Pathways

Distribute the **Pupil Resources** for **Queen Bee** to the children to help support them in their exploration of the different tasks. Use either of the **Curious Pupil Resources** depending on whether you are following the **Full Pathway** or the **Express Pathway**. The **Pupil Resources** for the remaining 5Cs are common to both pathways.

To provide additional flexibility in the classroom, each of the **Pupil Resources** is also provided in the **Teacher's Presentation** PowerPoint file (see slides 3–6), so you can display these at any time if you wish.

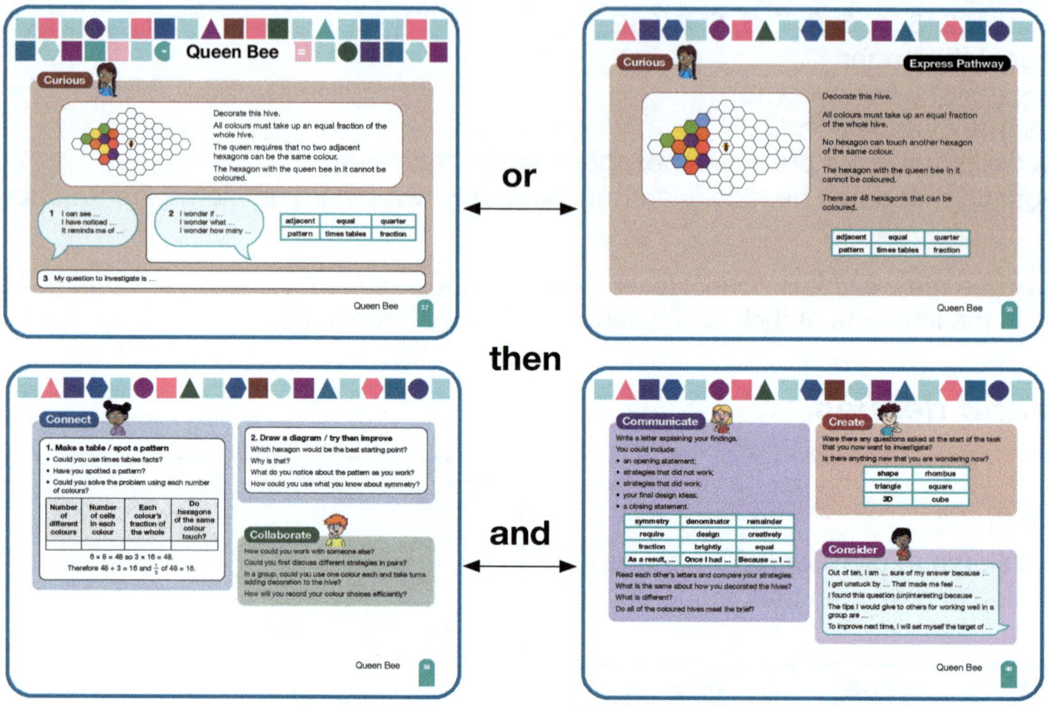

Express Pathway

If you are following the **Express Pathway**, spend a few minutes discussing the **Curious Pupil Resource** with the children. (This is slide 4, so you could display it if you wish.) Use some of the prompts on the previous page to help.

Two tasks involving different strategies are suggested for children to make connections.

1. Make a table / spot a pattern

Slide 7

Print slide 7 and distribute to children.

Encourage children to complete the table using multiplication and division facts to identify which colour combinations could be used. They may begin to notice a pattern that links the number of different colours to the fraction that would be taken up by each colour: the number of different colours used is the denominator of each colour's fraction of the whole, equivalent to the number of 48ths.

Number of different colours	Number of cells in each colour	Each colour's fraction of the whole	Do hexagons of the same colour touch?
1	48	$\frac{48}{48} = 1$	Yes
2	24	$\frac{24}{48} = \frac{1}{2}$	Yes
3	16	$\frac{16}{48} = \frac{1}{3}$	No
4	12	$\frac{12}{48} = \frac{1}{4}$	No
6	8	$\frac{8}{48} = \frac{1}{6}$	No
8	6	$\frac{6}{48} = \frac{1}{8}$	No
12	4	$\frac{4}{48} = \frac{1}{12}$	No
16	3	$\frac{3}{48} = \frac{1}{16}$	No
24	2	$\frac{2}{48} = \frac{1}{24}$	No
48	1	$\frac{1}{48}$	No

Slide 8

Print slide 8 and distribute to children.

Suggest that children use their printed copies for a 'trial then improve' approach to find out whether each number of different colours could result in no adjacent hexagons of the same colour.

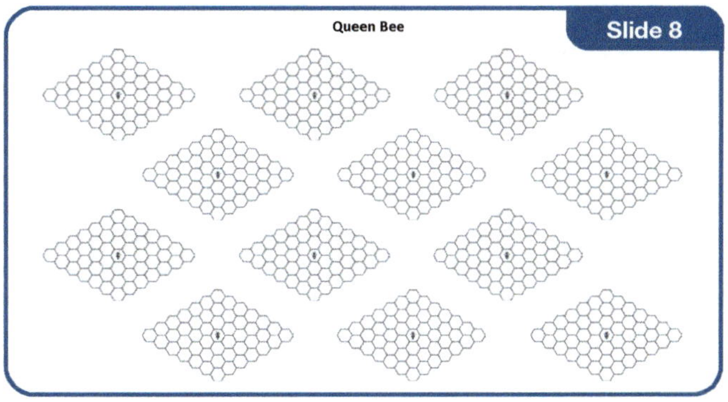

Slide 9

Show slide 9, if needed.

Encourage children to use known facts to work out unknown facts. Indicate the bar models on the slide, and note that children could use doubling and halving to create models such as these, which show how many of each colour there would be if six (for the first table) or three (for the second) different colours were used.

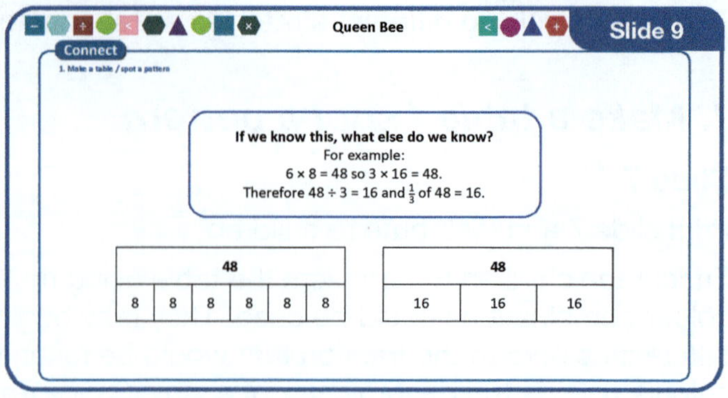

2. *Draw a diagram / try then improve*

Slide 10

Show slide 10. You may also wish to print and distribute this slide, or to provide more copies of slide 8, for children to colour.

Explain that one way of solving this problem is for children to choose a starting point using one colour and work outwards from that point.

- *Which hexagon would be the best starting point? Why is that?*

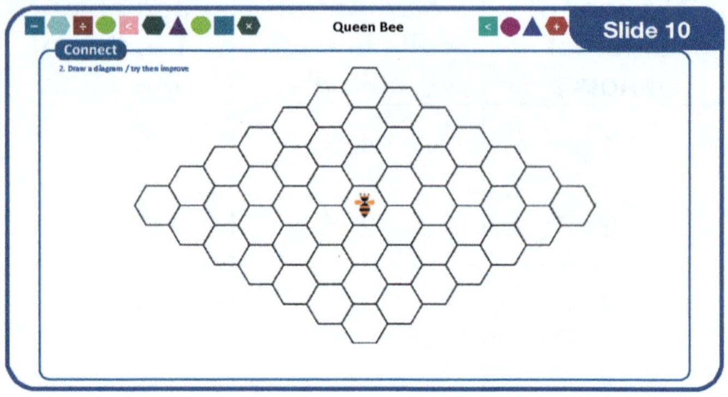

Children may not initially work systematically, and this may lead to more colours than necessary being used to decorate the hive. For example:

1.

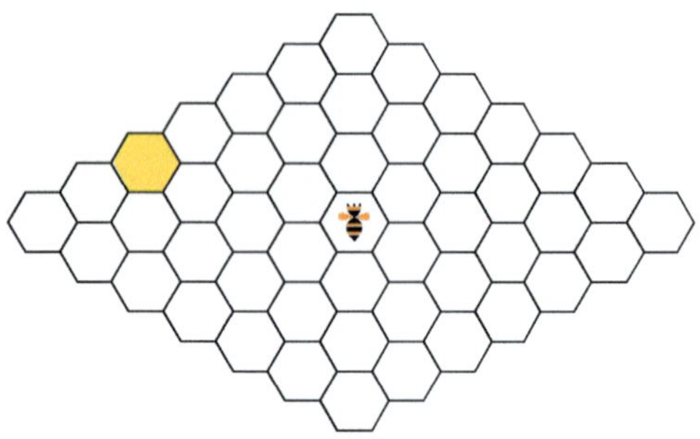

2.

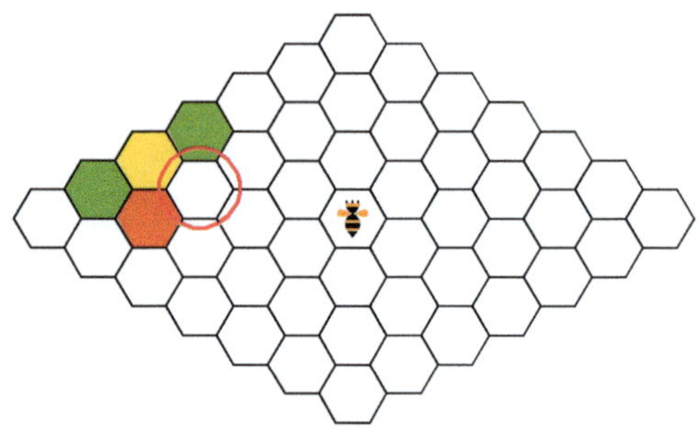

3.

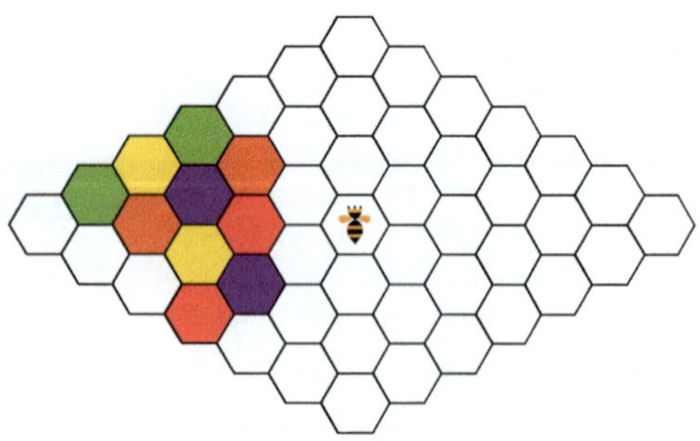

4.

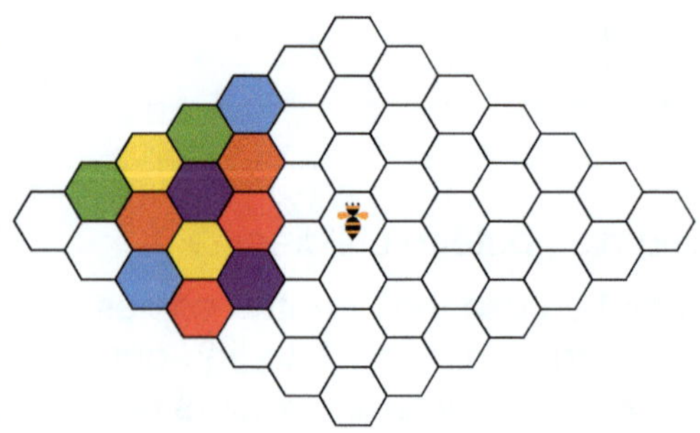

1. The pattern is begun with a yellow hexagon.
2. Different colours are used to work outwards from the original hexagon. Notice that none of the colours so far can be used for the next hexagon (indicated by the red ring).
3. A new colour is used for this hexagon (purple) and the pattern is continued. At this point, five colours have been used. If there should be equal fractions of each colour, five colours is not an option as 48 is not a multiple of five. Another colour is needed.
4. Blue has been added: six colours have now been used, and there are two hexagons of each colour. Systematic working, adding one hexagon of each colour at a time, will ensure that an equal fraction of each colour continues to be used.

To find the largest possible fraction of the total hexagons for one colour ($\frac{1}{3}$), children must work out the fewest possible number of colours that could be used (3).

After children have worked on the problem independently, use slide 10 to work through it systematically together.

Slide 11

Display slide 11, to prompt discussion about how using symmetry can speed up the decoration of the hive.

- *What do you notice about the pattern so far?*
- *How could you use what you know about symmetry?*

The children could apply these strategies to complete their hives.

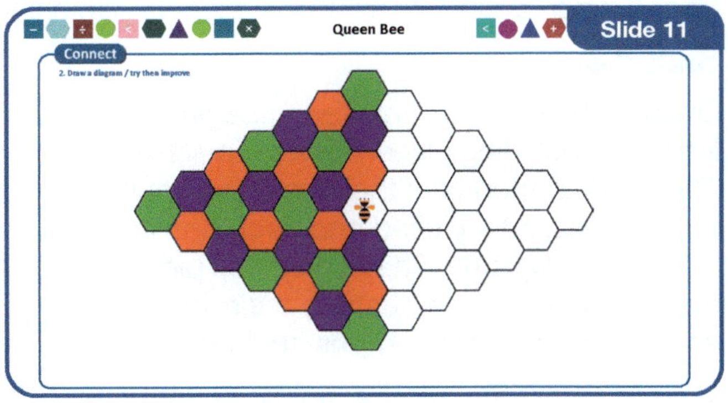

Slide 12

You may wish to display Slide 12 (the completed hive) for reference afterwards.

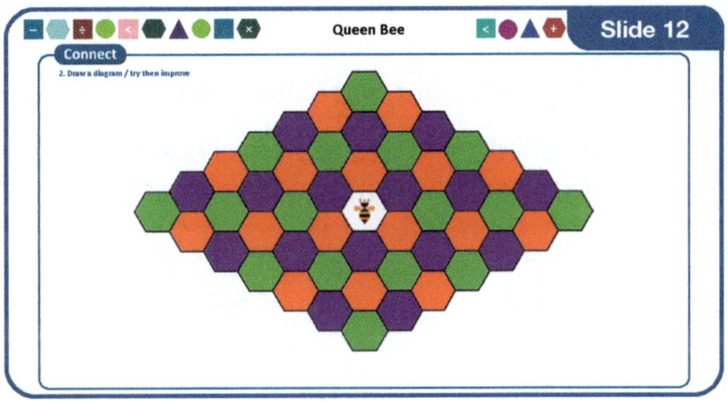

Connect

Refer to the **Collaborate** task of the **Pupil Resource**.

Children could communicate during the **Connect** task by discussing strategies in pairs before starting. Alternatively, they could work in groups, assigning one colour to each child, taking turns adding decoration to the hive and working together to decorate it.

Discuss time-efficient ways for children to record their colour choices.

Collaborate

Refer to the **Communicate** task of the **Pupil Resource**.

This task provides an opportunity for children to write structured and coherent explanations of their methods and results. Explain that they will write letters to other students who are trying to understand the task.

Recap letter-writing conventions such as beginning 'Dear …,' and ending by signing off.

Slide 13

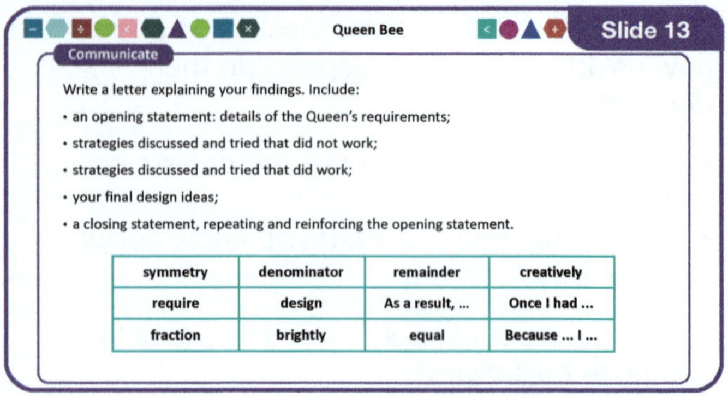

Display slide 13. Ask children to begin their letters by detailing Queen Bee's requirements. They should then take their readers through the processes they used and decisions they made, step by step. They could use pictures of the possible hives to illustrate their points or to show a process. Paragraphs could include:

- an opening statement: details of the Queen's requirements;
- strategies discussed and tried that did not work;
- strategies discussed and tried that did work;
- final design ideas;
- a closing statement, repeating and reinforcing the opening statement.

Encourage children to use the word bank as they write. Emphasise that their language needs to be precise and relatively formal to ensure accuracy and clarity.

symmetry	**denominator**	**remainder**
require	**design**	**creatively**
fraction	**brightly**	**equal**
As a result, …	**Once I had …**	**Because … I …**

Once their letters have been written, ask children to read each other's and compare their strategies and explanations.

- *What is the same about how you decorated the hives?*
- *What is different?*
- *Do all of the coloured hives meet the brief?*

Refer to the **Create** task of the **Pupil Resource**.

Slide 14

Display Slide 14.

- *Were there any questions asked at the start of the task that you now want to investigate?*
- *Is there anything new that you are wondering now?*

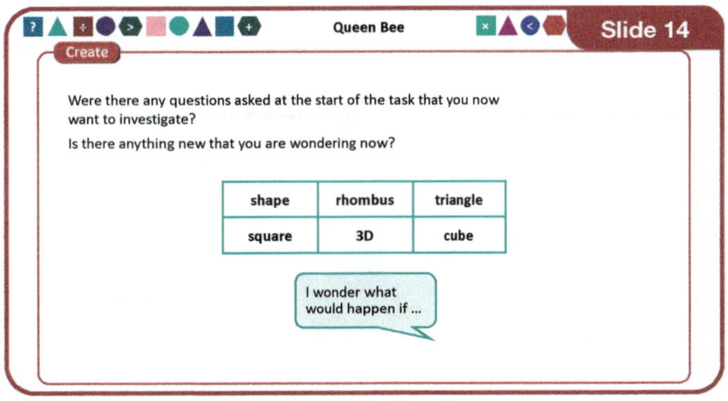

Children can use the word bank to prompt further questions of their own.

shape	**rhombus**	**triangle**
square	**3D**	**cube**

- I wonder what would happen if the hive was built from triangles or squares.
- I wonder what would happen if the hive still had 48 hexagons that could be coloured but was not a rhombus shape.
- I wonder what would happen if the hive was built out of cubes and the same rules applied for faces.

Refer to the **Consider** task of the **Pupil Resource**.

Slide 15

Display slide 15. Encourage children to think about the strategies that they have used. The prompts below may help to get them started.

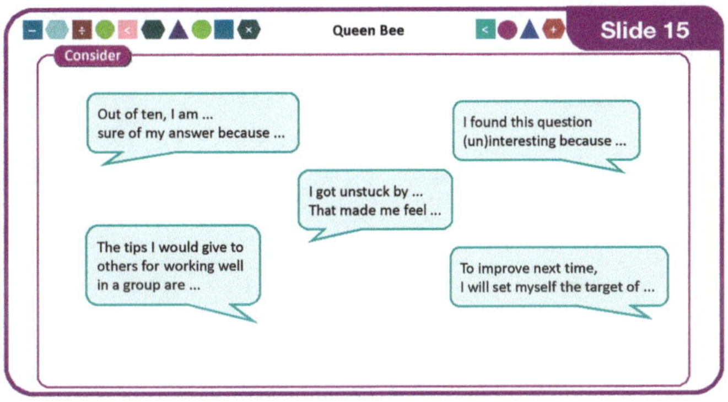

Cognitive learning domain

- Out of ten, I am ... sure of my answer because ...
- The new strategies I tried out during this task are …
- I learned …

Affective domain

- When I got stuck, I felt ...
- I got unstuck by ... That made me feel ...
- I said ‘aha’ when ... because ...
- I found this question (un)interesting because ...

Metacognitive domain

- The tips I would give to others for working well in a group are ...
- To improve next time, I will set myself the target of ...

Answers

Here are two possible solutions to the problem, using 3 and 4 colours:

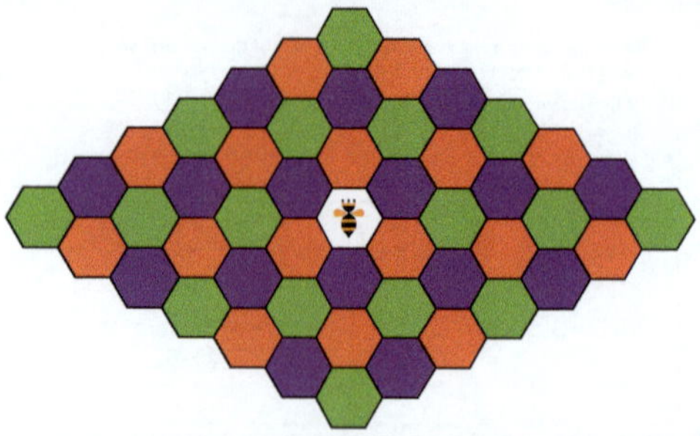

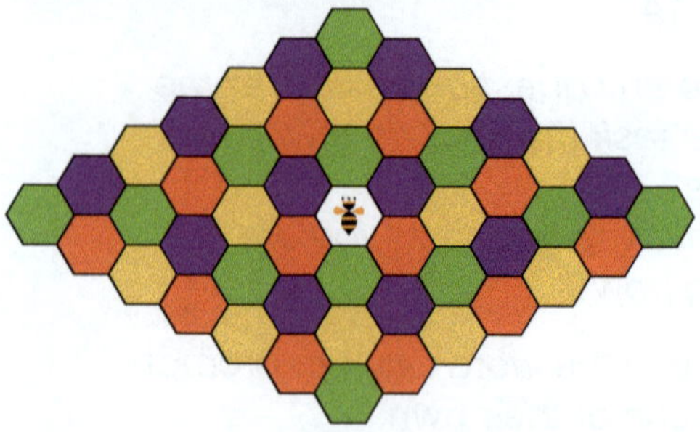

To achieve an equal number of hexagons in each colour group, children must take into account the total number of hexagons that can be coloured (48).

They may be able to recall multiplication facts to identify possible numbers of colours, such as 4×12 (4 colours or 12 colours), 6×8 (6 colours or 8 colours) or 2×24 (2 colours or 24 colours). They should also be able to extrapolate new facts from known facts, such as 3×16 (3 colours or 16 colours).

They may also use patterns and rules within times tables to eliminate numbers of colours. For example, 48 is not a multiple of 5 because it does not have 5 or a 0 as its ones digit.

These strategies will support with the first parameter of the task: that the total number of hexagons of each colour must be equal. However, children must refine these choices using the second rule. Using only one or two different colours will mean that the second rule (that no adjacent hexagons may be the same colour) cannot be met.

Balancing Act

Year 3

Introduction

Balancing Act uses images of different shapes on balance scales. Children use the mass of one shape to work out the masses of other shapes on other balance scales.

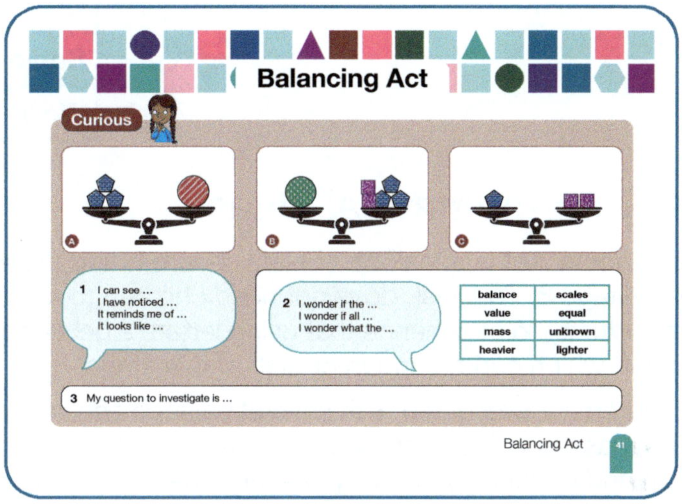

Children investigate similar problems, working with others to share ideas, and they create their own problems, all the while communicating their ideas and findings.

This task promotes greater-depth learning in a range of ways, supporting children to identify and articulate underlying mathematical structures. For example, in the **Curious** stage, greater depth is promoted by asking children to make rich mathematical contributions to discussions and suggest a variety of appropriate mathematical questions.

This **Teacher Guide** is supported by photocopiable **Pupil Resources** and **Teacher's Presentation** PowerPoint slides for whole-class or group display.

National Curriculum

Balancing Act covers these Year 3 National Curriculum objectives.

Measurement

- *The comparison of measures includes simple scaling by integers (for example, a given quantity or measure is twice as long or five times as high) and this connects to multiplication*. [non-statutory guidance]

Multiplication and division

- Solve problems, including missing number problems, involving multiplication and division, including positive integer scaling problems and correspondence problems in which *n* objects are connected to *m* objects.

The 6C cycle

The grid to the right shows the relative weighting given to each of the 6 Cs in **Balancing Act**.

Curious	★★★
Connect	★★
Collaborate	★
Communicate	★★
Create	★★
Consider	★★

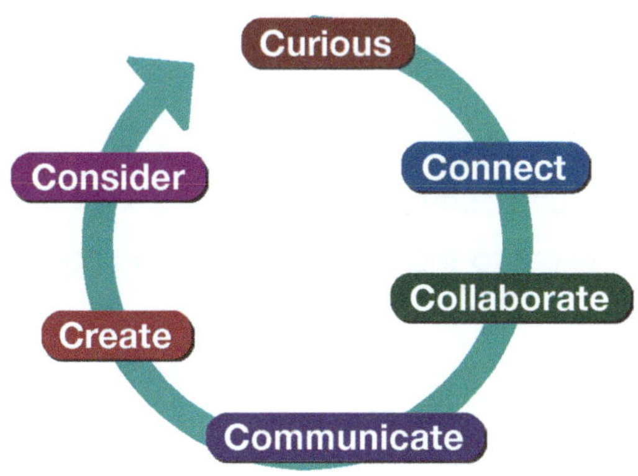

Problem-solving heuristics

The grid shows which problem-solving strategies children are likely to use when investigating the **Connect** tasks in **Balancing Act**. These strategies are further exemplified in the **Connect** stage below (pages 121–123).

Draw a diagram	✓
Make a table / spot a pattern	
Make a model / act it out	✓
Try then improve	
Create a systematic list	
Solve a simpler but related problem	✓
Work backwards	

Key mathematical ideas

Balancing Act is a task that builds secure foundations for later work on algebra in Year 6 and in Key Stage 3. This task does not need children to record algebraically, but it provides a conceptual opportunity to experience that using letters algebraically is simply short-hand for representing a model which can be drawn or built at this stage.

Children are presented with several balance scales with different shapes on each side of the balance. The masses of the shapes are not given. Children are encouraged to look at each balance scale in turn and describe what they see, and then to compare the different scales, focusing on what is the same and what is different.

Children might say or write statements such as:

- 3 pentagons are equal to 1 red striped circle. / 3 pentagons = 1 red striped circle (balance A)
- The green spotty circle is the same as 2 squares and 3 pentagons. / 1 green spotty circle = 2 squares + 3 pentagons (balance B)
- 1 pentagon weighs the same as 2 squares. / 1 pentagon = 2 squares (balance C)

They will develop a range of statements that describe relationships between the shapes. For example, 'the mass of the green spotty circle is the same as 2 squares and 1 red striped circle'.

When working at greater depth, children look for shortcuts and might write 'p' instead of 'pentagon' or 'r' instead of 'red circle', so look out for representations such as '3p = 1r'.

The mass of the red striped circle (42 kg) should be shared once the children have completed **Develop and deepen**. Once they know the mass of one shape, they can work out the mass of other shapes.

Using the initial statement '3 pentagons are equal to 1 red striped circle', children might now say: 'I can divide 42 kg by 3 to find the mass of a pentagon because the mass of 3 pentagons is 42 kg.'

3 pentagons = 42, so 1 pentagon = 42 ÷ 3 = 14 kg

They can then use this value to work out the mass of the other shapes:

1 pentagon = 2 squares (from slide 3)
14 = 2 squares, so 1 square = 14 ÷ 2 = 7 kg

and

1 green spotty circle = 2 squares + 3 pentagons (from balance B)
1 green spotty circle = 2 × 7 + 3 × 14 = 14 + 42 = 56 kg.

Answers to the tasks are given on page 125.

Teaching guidance

If you are using the **Express Pathway**, please turn to pages 121–125.

Teaching guidance for the **Full 6C Pathway** is given below.

The **Teacher's Presentation** PowerPoint file has slides specifically mentioned in this teaching guidance. Some of the slides include notes (within the PowerPoint file) providing 'stumble support' suggestions, that may help you to support children in their problem-solving thinking and enable access for children who are less confident.

Awe and wonder

Reveal slides 1–3 one at a time, allowing time for discussion and observations as prompted below.

Slide 1

- I can see three pentagons and a circle.
- I have noticed that the circle is larger than a pentagon.
- It reminds me of a seesaw in the park.
- It looks like the shapes are equal in mass if it is a balancing scale.
- 3 pentagons are the same / have the same mass as a circle.
- It reminds me of a number sentence: 3 pentagons = 1 red circle.

⚠ Watch out for children who use 'weight' instead of 'mass'. Weight/mass may be used interchangeably in KS1. In KS2 it is important that children use mass (kg/g).

Slide 2

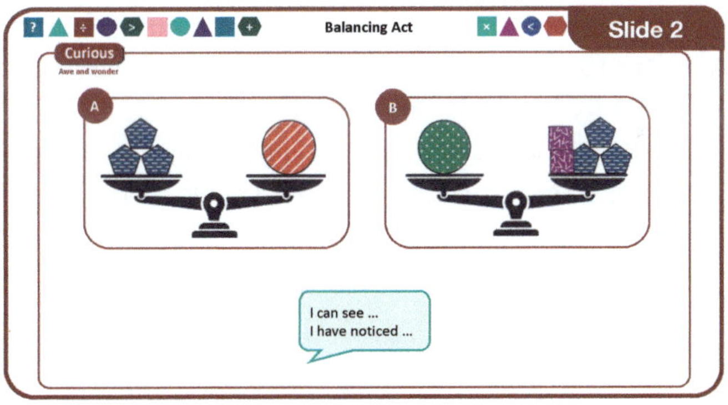

- I can now see a circle, 2 squares and the 3 pentagons.
- I have noticed that the number of pentagons is the same in both pictures.
- I have noticed that the circles are different colours.

Slide 3

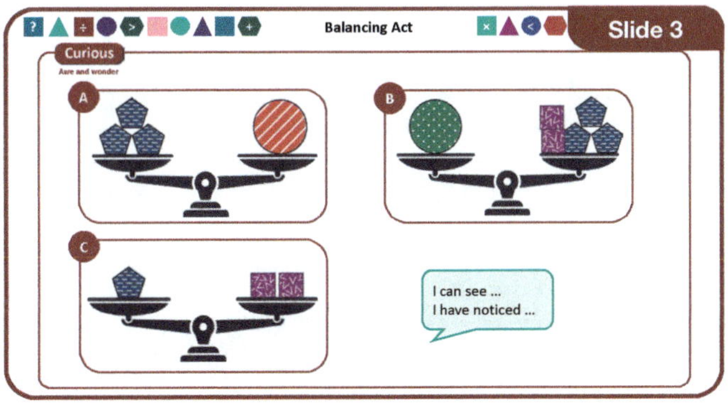

- I have noticed that the pentagon appears again.
- I can see that 2 squares are equal to a pentagon.

Curious

Develop and deepen

Slide 4

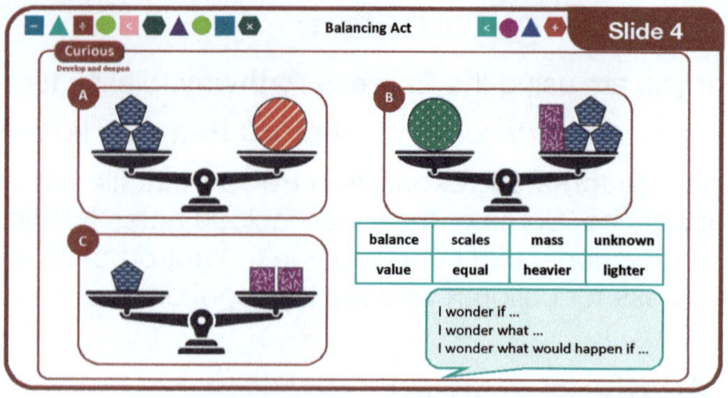

Display slide 4. Encourage children to use the words in the word bank to develop and deepen their thoughts and questions.

Possible lines of enquiry

- I wonder if all of the shapes represent a different mass.
- I wonder what the mass of the red striped circle is.

balance	**scales**
value	**equal**
mass	**unknown**
heavier	**lighter**

Possible lines of enquiry linked to the word bank:

- I wonder if the shapes have the same value in each picture.
- I wonder if the red striped circle is lighter than the green spotty circle as there are 2 squares with the 3 pentagons in the second balance.

Possible lines of enquiry that children may raise now, but should be held back until the **Create** task:

- I wonder what would happen to the values if one of the balances wasn't equal.

Consider whether / how to collate and display all the suggested questions, and whether / how to steer children towards selecting an appropriately challenging question.

When children have developed their questions, reveal that the red striped circle weighs 42 kg.

Pupil Resources for Full and Express Pathways

Distribute the **Pupil Resources** for **Balancing Act** to the children to help support them in their exploration of the different tasks. Use either of the **Curious Pupil Resources** depending on whether you are following the **Full Pathway** or the **Express Pathway**. The **Pupil Resources** for the remaining 5Cs are common to both pathways.

To provide additional flexibility in the classroom, each of the **Pupil Resources** is also provided in the **Teacher's Presentation** PowerPoint file (see slides 5–8), so you can display these at any time if you wish.

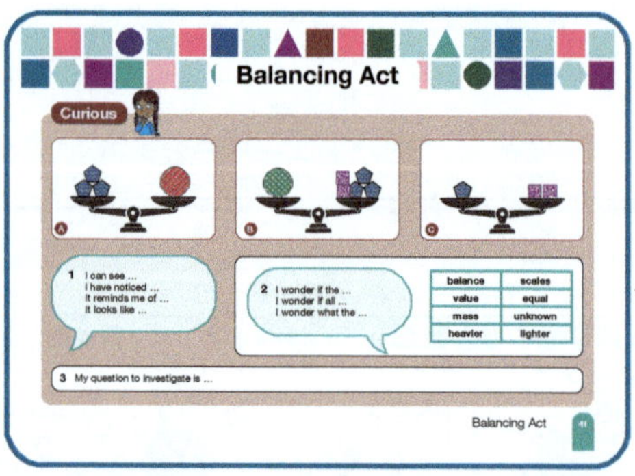

or

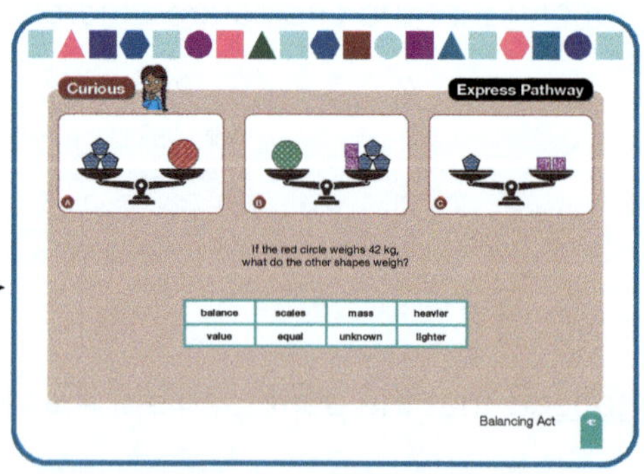

then

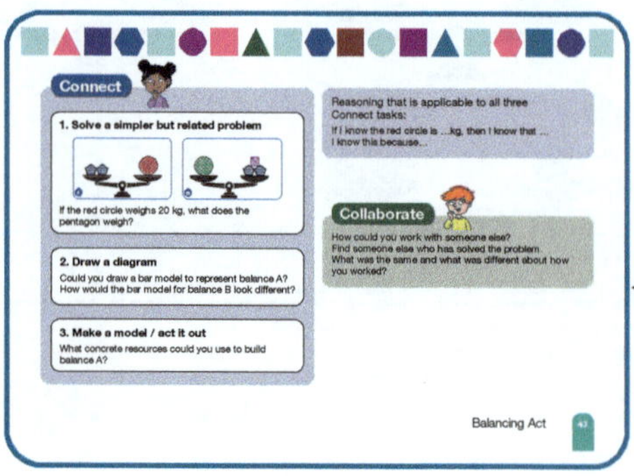

and

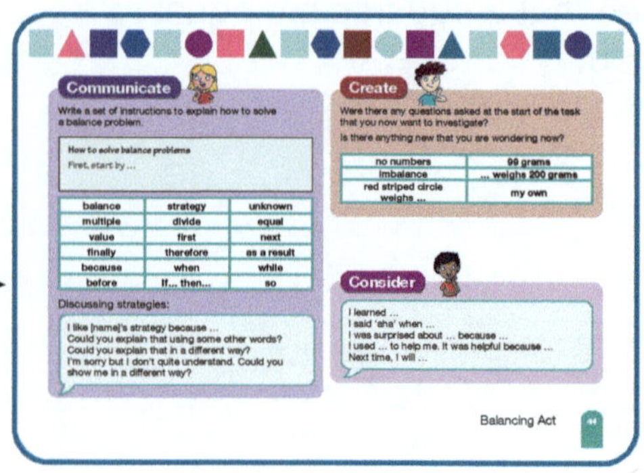

Express Pathway

If you are following the **Express Pathway**, spend a few minutes discussing the **Curious Pupil Resource** with the children. (This is slide 6, so you could display it if you wish.) Use some of the prompts on the previous page to help.

Three tasks involving different strategies are suggested for children to make connections.

1. Solve a simpler but related problem

Slide 9

Display slide 9 (balances D and E).

Focus children's attention on balance D.

- *If the mass of the red striped circle is 20 kg, what is the mass of the pentagon?*
- If I know the red striped circle is … kg, then I know that … because
- *Can you draw a diagram to prove your answers?*

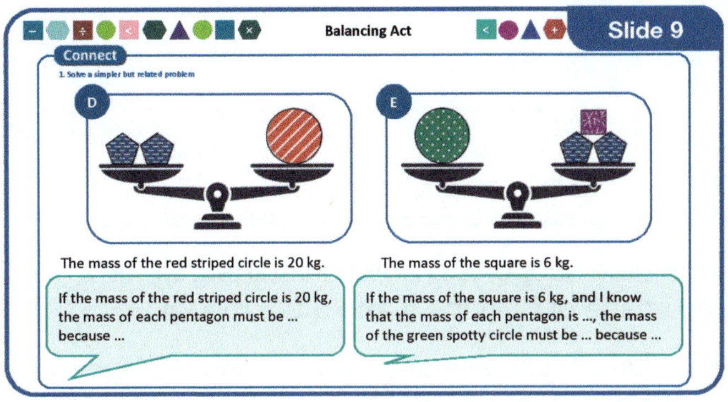

Now, look at balance E.

- *What is different this time?* (From balance D, now we know the masses of the pentagons.)

⚠ Watch out for children who think that the circles have to have the same value in each diagram.

2. Draw a diagram

Bar models provide a clear visual aid to allow children to compare the size of each shape and see how to find an unknown mass or value. For example, for balance A, if the mass of the red striped circle is known, then to find a pentagon, this needs to be divided by 3. As more unknowns are discovered, encourage the children to annotate their diagrams. Once the mass of the pentagon is known, this unlocks the mass of all of the other shapes.

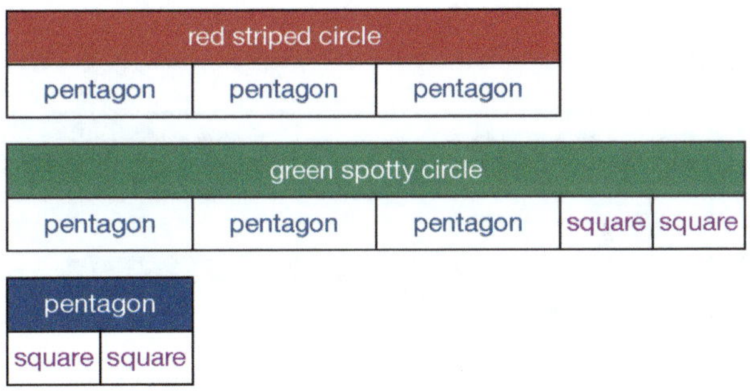

- *Have you shown that all the pentagons are the same value in your bar models? Make sure they are all the same size.*
- *If we know the value of the red striped circle, how can we find the value of a pentagon?*
- *Where else can we put this information on our diagrams?*
- *If we know this, what else do we now know?*

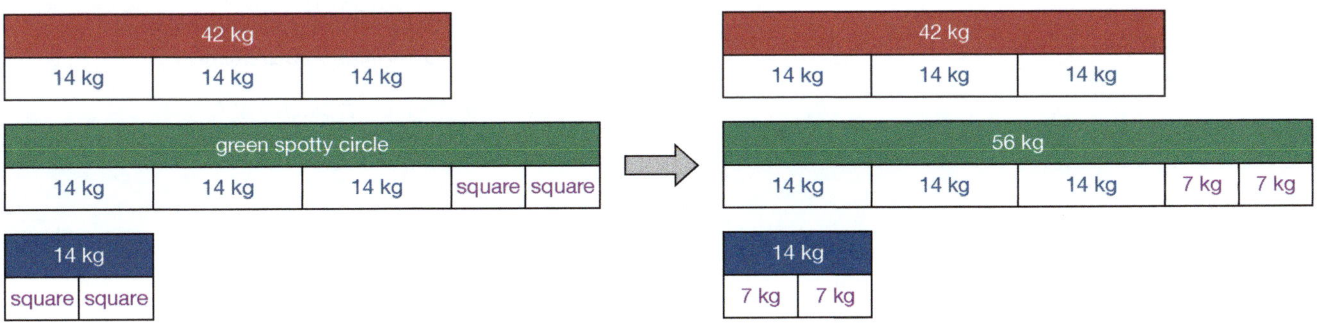

Connect

For additional support, slides 10–13 can be used to explore how to turn balances A, B and C into bar models. If possible, once a slide has been explored, make a display copy of the bar model and balance (for example, on a flipchart or working wall), so that children can see how the different bars show the relationships.

Slide 10

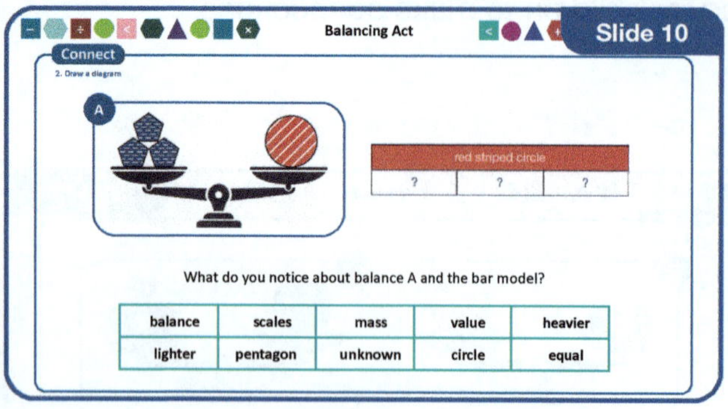

This slide supports children in starting to turn the image of balance A into a bar model.

- *What do you see and notice about balance A and the bar model?*

Slide 11

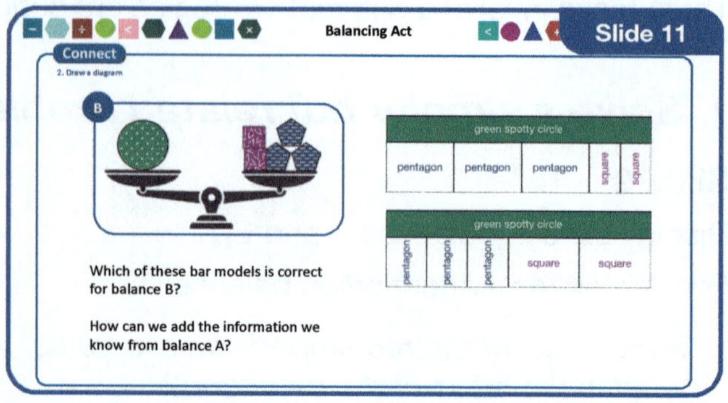

Before showing slide 11, ask the children if they could draw what they think the bar model for balance B might look like.

There is no single correct answer at this stage as the mass of the squares and the green spotty circle are unknown.

⚠ Check the children understand that at this stage there is more than one correct bar model.

Slide 12

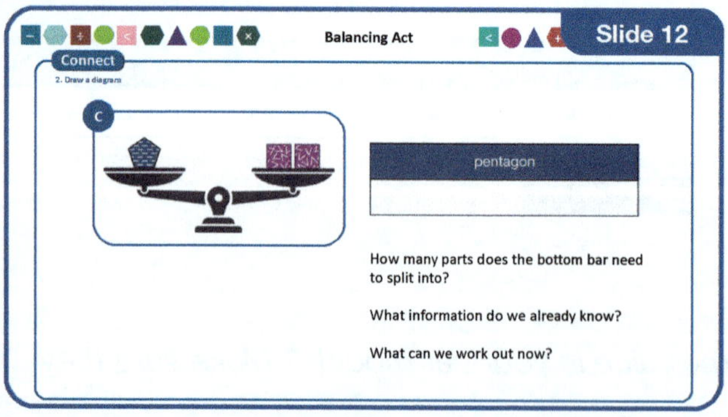

Before showing slide 12, ask the children if they could draw what they think the bar model for balance C might look like.

- *How many parts does the other bar have to split into? Why?*

Slide 13

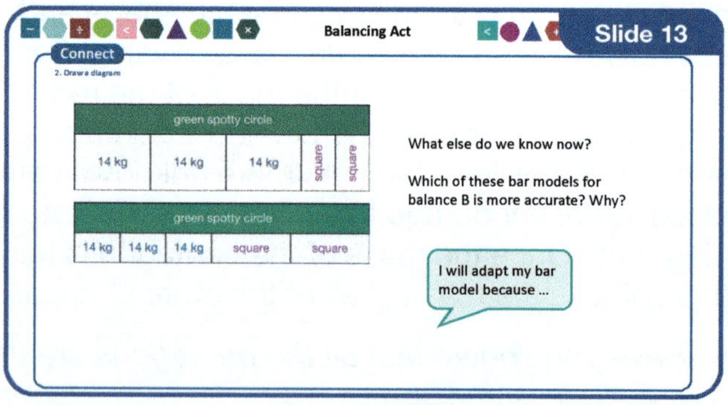

This slide allows children to see why it is important to update the bar models with new information as it is obtained. It also makes children think more deeply about how we represent relationships accurately.

- *What else do we know now?*

3. *Make a model / act it out*

Additional resources required: number rods (alternatively, balance scales and masses)

This strategy will be easier to implement if children have already done the task 'Solve a simpler but related problem'. Using concrete resources that are proportional to each other, children act out the scenarios. For example, number rods could be used to build the bar models.

Slide 14

When displaying slide 14, it is important that actual concrete resources are used to make the model; do not simply use the images on the slide.

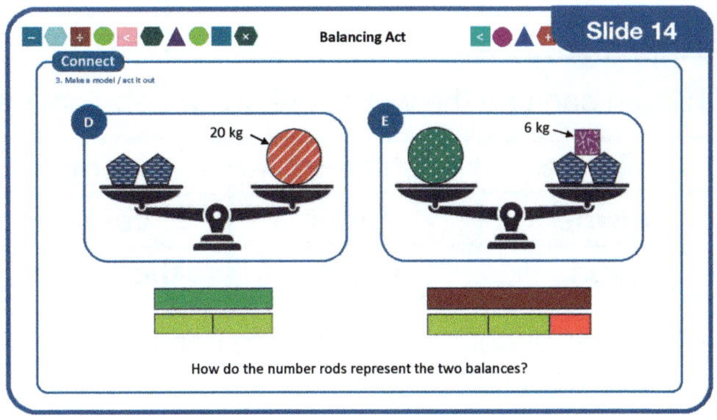

Balance D

- *What is the same and what is different between the model and the picture of balance D?*

Balance E

- *What is the same and what is different between this model and the model representing balance D?*
- *How could you show balances D and E using bar models?*

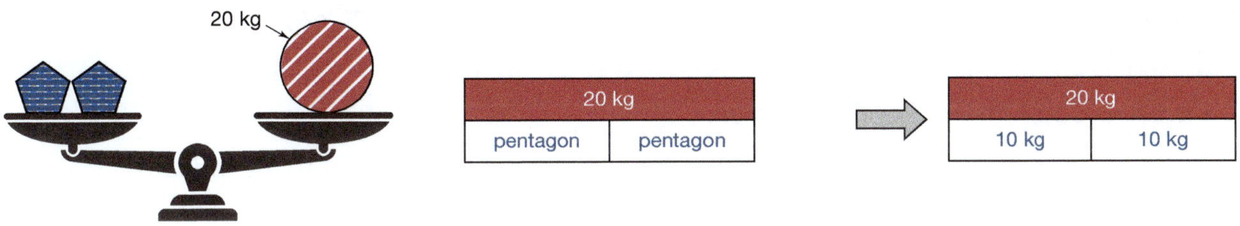

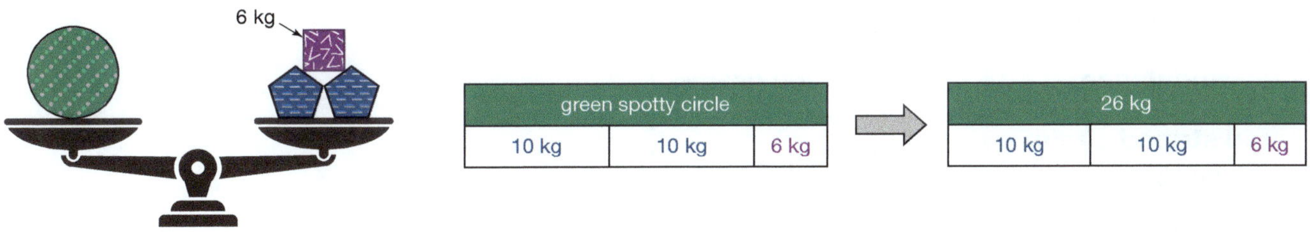

Alternatively, children could use balance scales and masses to re-enact the diagrams.

Refer to the **Collaborate** task of the **Pupil Resource**.

Pairs of children who worked on the same task in the **Connect** stage could compare how they solved the problem.

- *What was the same? What was different?*

Communicate

Refer to the **Communicate** task of the **Pupil Resource**.

This task provides an opportunity for the children to write instructions.

> How to solve balance problems
>
> First, start by ...

Children can use the word bank for vocabulary support.

balance	**strategy**	**unknown**	**multiple**	**because**	**when**
divide	**equal**	**value**	**first**	**while**	**before**
next	**finally**	**therefore**	**as a result**	**if ... then ...**	**so**

Afterwards, encourage children to compare the strategies that they have used, and ask each other questions for clarification.

- I like [name]'s strategy because …
- Could you explain that in some other words?
- Could you explain that in a different way?
- I'm sorry but I don't quite understand. Could you show me in a different way?

Create

Refer to the **Create** task of the **Pupil Resource**.

- *Were there any questions asked at the start of the task that you now want to investigate?*
- *Is there anything new that you are wondering now?*

Children can use the word bank to prompt further questions of their own.

no numbers	**99 grams**
imbalance	**… weighs 200 grams**
red striped circle weighs …	**my own …**

- I wonder what would happen to the values if one of the balances wasn't equal.
- I wonder if it would work if no numbers were involved.
- I wonder what would happen if the red striped circle wasn't 42 kg but was 99 grams.
- I wonder if I could start with a different shape. For example, what if the square weighed 200 grams?
- I wonder if I can design my own puzzle using these shapes and balances.

Slide 15

If needed, display slide 15.

- *What's the same and what's different in these diagrams compared to balances A–C?*
- This time one of the balances is not equal.
- The two squares must be lighter than a pentagon rather than equal to it.

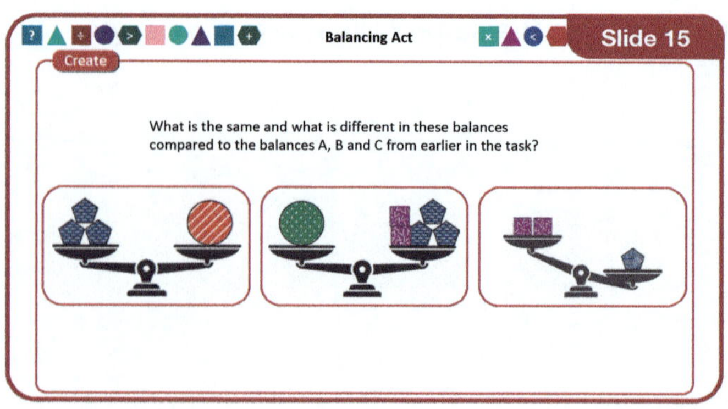

Refer to the **Consider** task of the **Pupil Resource**.

Encourage children to think about the strategies that they have used and also the instruction writing of the other children.

The prompts below may help get children started.

Cognitive learning domain

- The trickiest part of solving my question was …
- I learned …
- The best strategy was …

Affective domain

- The fun bit of my learning was … It was fun because …
- I said 'aha' when … I said it because …
- I was surprised about … because …

Metacognitive domain

- I got stuck when … I unstuck myself by …
- I used … to help me. It was helpful because …
- Next time, I will …

Answers

The three balance scale images can be represented like this:

3 pentagons = 1 red striped circle
1 green spotty circle = 2 squares + 3 pentagons
1 pentagon = 2 squares

Substituting the mass given for the red striped circle, 42 kg, gives:

[using balance A] 3 pentagons = 42 kg, so 1 pentagon = 42 kg ÷ 3 = 14 kg

Therefore:

[using balance C] 14 kg = 2 squares, so 1 square = 14 kg ÷ 2 = 7 kg

And so:

[using balance B] 1 green spotty circle = 2 × 7 kg + 3 × 14 kg = 14 kg + 42 kg = 56 kg

Connect: Solve a simpler but related problem

The balances on slide 9 can be represented as:

2 pentagons = 20 kg
1 green spotty circle = 2 pentagons + 6 kg

Working out the mass of 1 pentagon:

2 pentagons = 20 kg, so 1 pentagon = 20 kg ÷ 2 = 10 kg

And therefore:

1 green spotty circle = 2 × 10 kg + 6 kg = 26 kg

Create

The unbalanced scale in slide 15 can be represented using the inequality sign:

1 pentagon > 2 squares

Perfect Poggle Pens

Year 4

Introduction

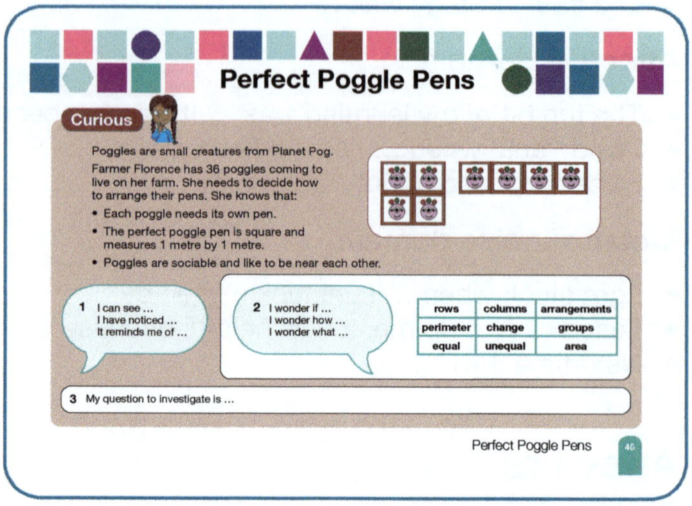

Perfect Poggle Pens gives children an opportunity to explore perimeter and area by considering the best ways to arrange pens for 36 imaginary creatures called 'poggles'. Further challenge is included by suggesting additional criteria to be considered when creating their arrangements. Children should make connections with their knowledge of multiplication and division facts, as well as their experience with remainders.

Children can use a range of heuristics to justify which arrangement they consider the most effective, then evaluate the models and diagrams used by others. Children may encounter proofs that contradict their thinking, hence seeing that more than one answer is possible. Teachers can choose, if appropriate, to encourage children to use what they have found out to ensure they have found all possible answers.

This task promotes greater-depth learning in a range of ways, supporting children to identify and articulate underlying mathematical structures. For example, in the **Communicate** stage, greater depth is promoted by asking children to develop clear, accurate and well-organised communication, using correct mathematical language and illustrating key ideas using diagrams or models.

This **Teacher Guide** is supported by photocopiable **Pupil Resources** and **Teacher's Presentation** PowerPoint slides for whole-class or group display.

National Curriculum

Perfect Poggle Pens covers these Year 4 National Curriculum objectives.

Measurement

- Measure and calculate the perimeter of a rectilinear figure (including squares) in centimetres and metres.
- Find the area of rectilinear shapes by counting squares.

Multiplication and division

- Recognise and use factor pairs and commutativity in mental calculations.

The 6C cycle

The grid below shows the relative weighting given to each of the 6 Cs in **Perfect Poggle Pens**.

Curious	★★
Connect	★★★
Collaborate	★★
Communicate	★★★
Create	★★
Consider	★

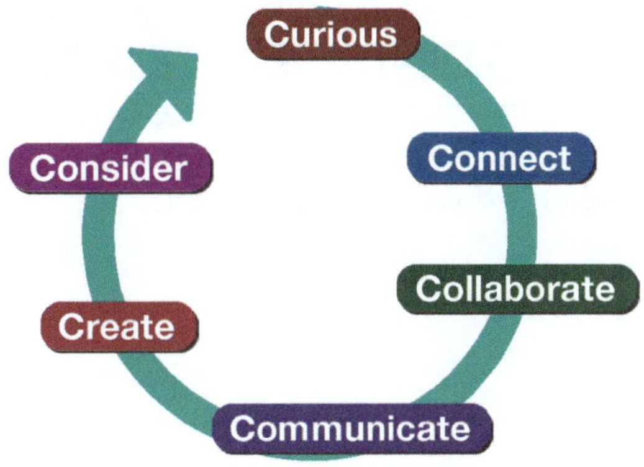

Problem-solving heuristics

The grid shows which problem-solving strategies children are likely to use when investigating the Connect tasks in **Perfect Poggle Pens**. These strategies are further exemplified in the Connect stage below (pages 130–132).

Draw a diagram	✓
Make a table / spot a pattern	✓
Make a model / act it out	✓
Try then improve	✓
Create a systematic list	✓
Solve a simpler but related problem	
Work backwards	✓

Key mathematical ideas

Children should use their knowledge of equal groups, multiplication and division facts and factors to find all possible arrangements for the 36 poggles.

In Year 4, area is calculated by counting squares within rectilinear shapes, not by using the formula of length $\times$ width. However, some children may make this link between factor pairs and the number of squares, and can use this to find all possible solutions.

Factor pairs of 36: 36×1 18×2 12×3 9×4 6×6

Children may notice relationships between the factor pairs and use multiplicative laws to find all pairs. For example, using the doubling and halving multiplicative relationship, it can be proven that $18 \times 2 = 9 \times 4$.

Ensure that children consider the commutative properties of multiplication, for example, by working out that 9 rows of 4 pens is a solution, as well as 4 rows of 9 pens.

If choosing to use this task in Year 3, adapt the problem to omit references to area. Children may, however, discuss the number of squares, recognising that at first there are 4 poggles and 4 squares within each arrangement and then, when exploring 36 poggles, notice the relationship between the 36 poggles and the 36 squares.

Answers to the tasks are given in situ, and on page 134.

Teaching guidance

If you are using the **Express Pathway**, please turn to pages 129–134.

Teaching guidance for the **Full 6C Pathway** is given below.

The **Teacher's Presentation** PowerPoint file has slides specifically mentioned in this teaching guidance. Some of the slides include notes (within the PowerPoint file) providing 'stumble support' suggestions, that may help you to support children in their problem-solving thinking and enable access for children who are less confident.

Awe and wonder

Slide 1

Display slide 1.

Allow children to discuss their initial thoughts.

- I can see that there are 8 poggles altogether.
- I can see that there are 4 poggles in each set.
- I have noticed that in one set the poggles are in a square and in the other set the poggles are in a line / make a rectangle.
- I have noticed that the poggles each have their own square.
- I have noticed that the poggles are all touching in the square but not in the rectangle.
- It reminds me of arrays: 2 groups of 2, and 1 group of 4.

Slide 2

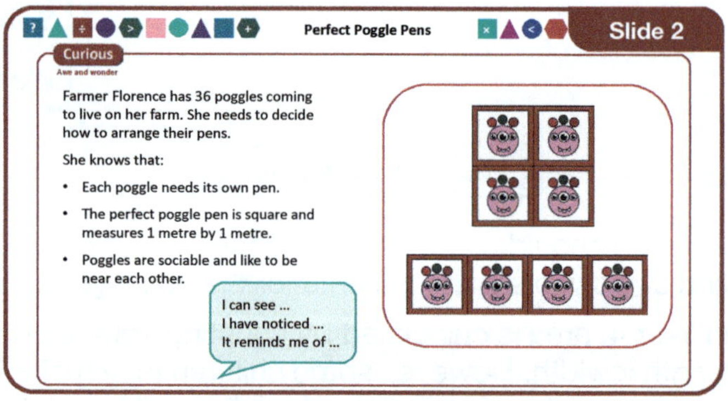

Display slide 2.

- I can see that the poggles like to be near each other, so they might like the square arrangement better.

Develop and deepen

Slide 3

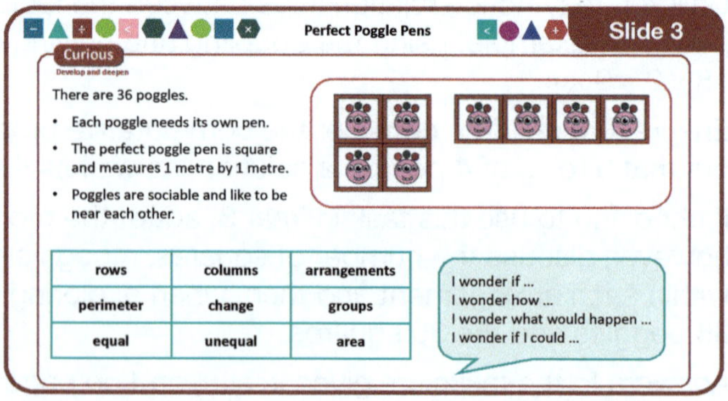

Display slide 3.

Encourage children to use the words in the word bank to develop and deepen their thoughts and questions.

rows	columns	arrangements
perimeter	change	groups
equal	unequal	area

Possible lines of enquiry linked to the word bank:

- I wonder if all of the rows and columns need to have an equal number of poggles in them.
- I wonder if the perimeter of my arrangement will change.
- I wonder how the perimeter of my arrangement will change each time.
- I wonder if the area of my arrangement will change.

Possible lines of enquiry that children may raise now, but should be held back until the **Create** task:

- I wonder if I had fewer poggles, would I be able to make fewer arrangements.
- I wonder what would happen if the rows could have unequal groups. Could I still make the same number of arrangements?
- I wonder if I could make arrangements where the sides don't have to meet. How many could I make?
- I wonder what would happen if each poggle needed 2 metres by 2 metres.

Consider whether / how to collate and display all the suggested questions, and whether / how to steer children towards selecting an appropriately challenging question.

Pupil Resources for Full and Express Pathways

Distribute the **Pupil Resources** for **Perfect Poggle Pens** to the children to help support them in their exploration of the different tasks. Use either of the **Curious Pupil Resources** depending on whether you are following the **Full Pathway** or the **Express Pathway**. The **Pupil Resources** for the remaining 5Cs are common to both pathways.

To provide additional flexibility in the classroom, each of the **Pupil Resources** is also provided in the **Teacher's Presentation** PowerPoint file (see slides 4–7), so you can display these at any time if you wish.

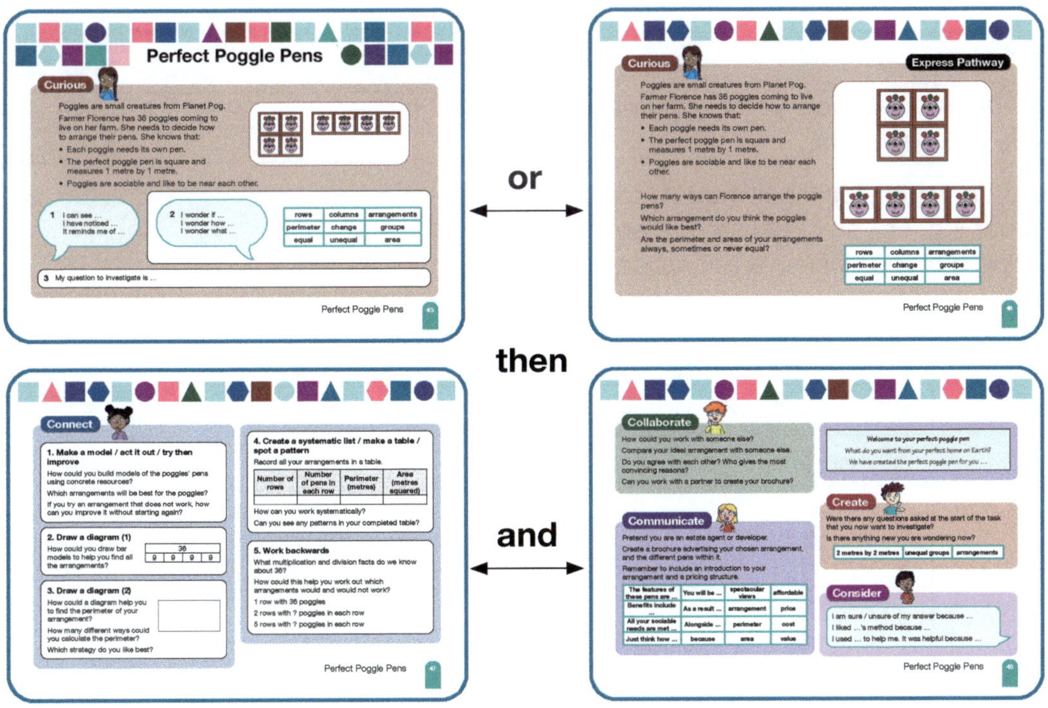

Express Pathway

If you are following the **Express Pathway**, spend a few minutes discussing the **Curious Pupil Resource** with the children. (This is slide 5, so you could display it if you wish.) Use some of the prompts on the previous page to help.

Five tasks involving different strategies are suggested for children to make connections.

1. Make a model / act it out / try then improve

Additional resources required: interlocking cubes (or other concrete resources that can be used to build arrays)

Slide 8

Display slide 8.

- *If you use one cube to represent each pen, how many cubes will you need in total?*

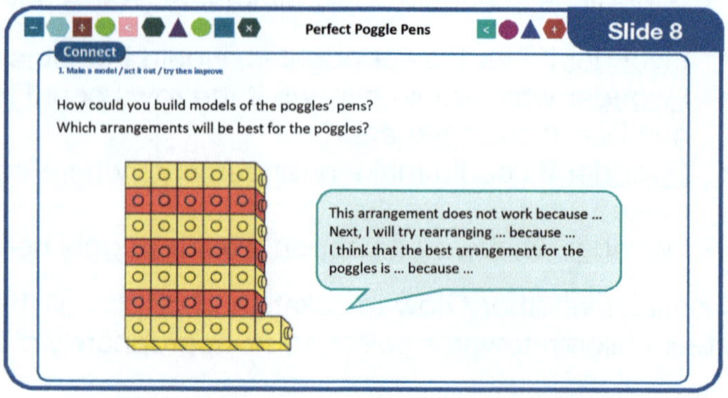

Ask children to replicate the model shown on slide 8.

- *Why is this not the best arrangement?*

Ask children to come up with their own arrangements.

- *If this arrangement doesn't work, how can you improve it without starting again?*

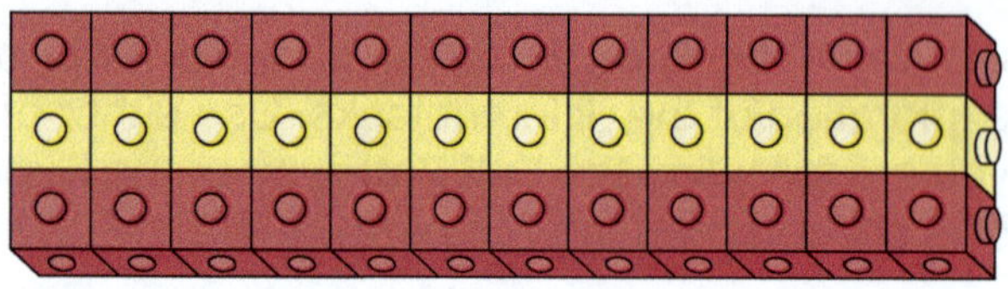

Children should be aiming to make models that are rectangular (for example, a 3 by 12 array).

2. Draw a diagram (1)

Slide 9

Display slide 9.

- *How could bar models help you to find all the possible arrangements?*

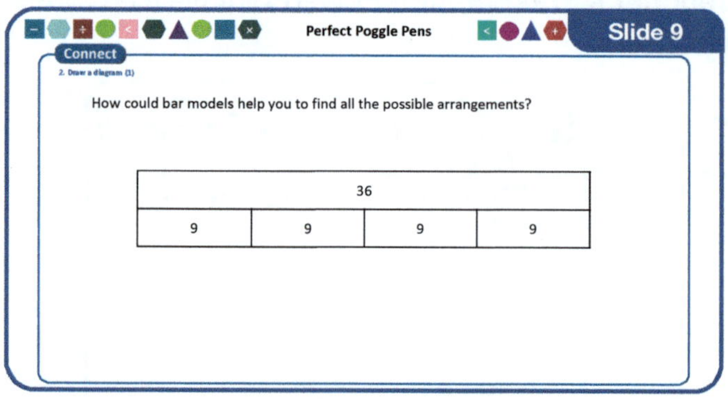

Children could draw individual bar models for each equal grouping, or you could steer them towards creating a comprehensive bar model like this:

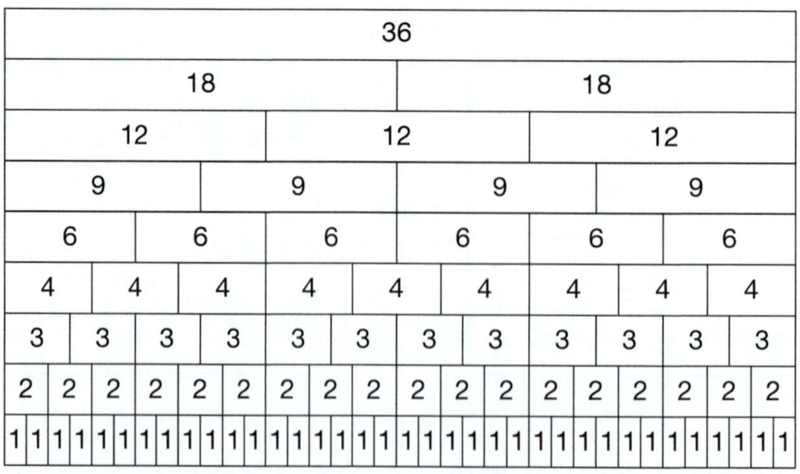

Encourage children to work through the factor pairs systematically.

- *Can you see any patterns when the possible groupings are shown in order?*
- <u>I have noticed</u> that the group sizes are decreasing but the number of groups are increasing.

3. *Draw a diagram (2)*

Additional resources required: squared paper (optional)

Slide 10

Display slide 10.

- *How could a diagram help you to find the perimeter of your arrangement?*

Children could draw rectangles to represent their arrangement, either on squared paper (using 1 square to represent 1 poggle pen), or on plain paper (recording the dimensions along the sides).

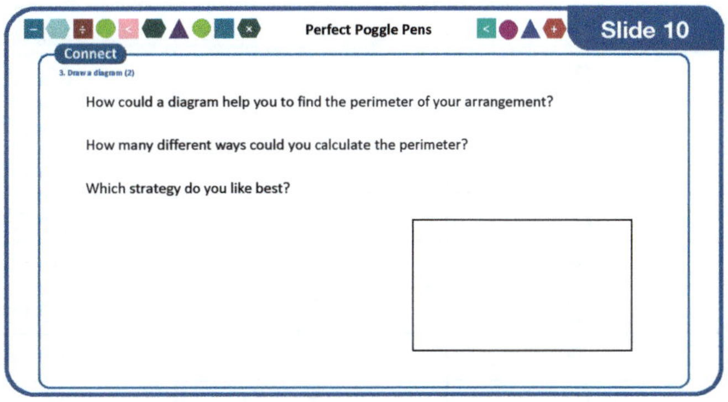

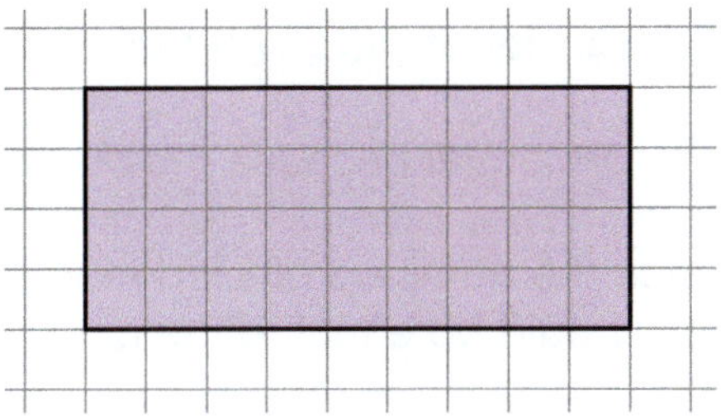

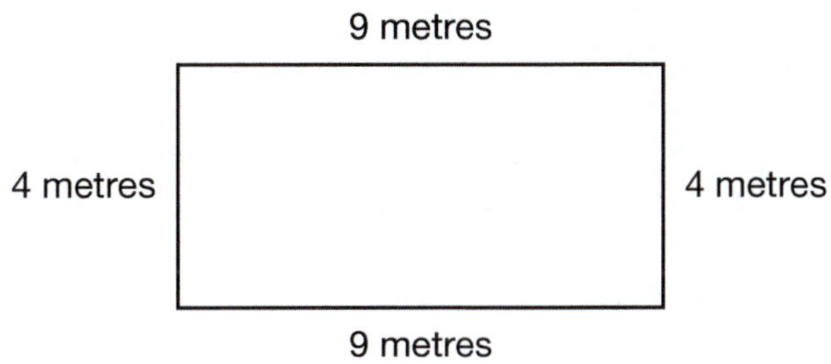

- *How many different ways could you calculate the perimeter?*

For example, the perimeter could be calculated by adding together each of the four sides or by adding a width and length together, and then doubling.

- *Which strategy do you like best?*

⚠ Watch out for children who do not mark their starting point when counting the perimeter and therefore count too many or too few outside edges.

4. *Create a systematic list / make a table / spot a pattern*

Slide 11

Print slide 11 and distribute to the children.

Encourage children to think systematically about their arrangements to ensure that they have considered all possibilities. This will also help them to spot any patterns between their arrangements.

For children who are not working with area, ask them to cross out the final column heading.

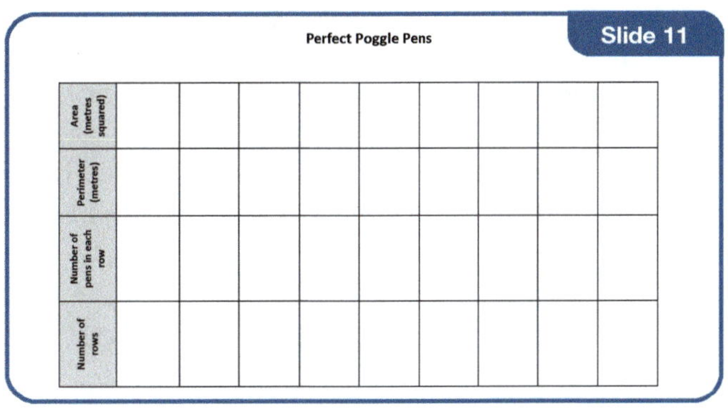

⚠ For children who do choose to work with area, ensure they understand and can describe the difference between area and perimeter. Highlight that the units of measurement are metres for perimeter (abbreviated to m) and metres squared for area (abbreviated to m^2).

Connect

Number of rows	Number of pens in each row	Perimeter (metres)	Area (metres squared)
1	36	74 m	36 m^2
2	18	40 m	36 m^2
3	12	30 m	36 m^2
4	9	26 m	36 m^2
6	6	24 m	36 m^2
9	4	26 m	36 m^2
12	3	30 m	36 m^2
18	2	40 m	36 m^2
36	1	74 m	36 m^2

- <u>I have noticed</u> that the greater the difference between the number of rows and the number of pens in each row, the greater the perimeter is.
- <u>I have noticed</u> that some of the perimeters are equal. They are equal because the number of rows and number of pens in each row have swapped.
- <u>I realised that</u> I did not need to calculate the perimeter once the numbers started reappearing. So, for the perimeter of 12 rows of 3 pens, I knew that this would be equal to the perimeter of 3 rows of 12 pens.
- <u>I have noticed</u> that the area is always 36 cm squared. I think that this is because we are always arranging 36 pens and so no matter the arrangement, the area will not change.

5. Work backwards

Slide 12

Display slide 12.

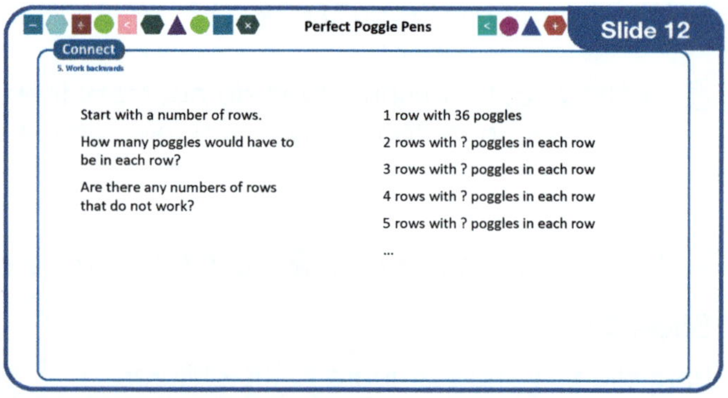

- *What multiplication and division facts do we know about 36?*
- *How could this help us to work out which arrangements would and would not work?*
- *Start with a number of rows.*
- *How many poggles would have to be in each row?*
- *Are there any numbers of rows that do not work?*
- There can't be 5 rows, because 5 is not a multiple of 36. If we had 5 rows, there couldn't be an equal number of poggles in each row, and one poggle would only have one neighbour.

36 ÷ 1	1 row with 36 poggles	36 ÷ 6	6 rows with 6 poggles in each row
36 ÷ 2	2 rows with 18 poggles in each row	36 ÷ 9	9 rows with 4 poggles in each row
36 ÷ 3	3 rows with 12 poggles in each row	36 ÷ 12	12 rows with 3 poggles in each row
36 ÷ 4	4 rows with 9 poggles in each row	36 ÷ 18	18 rows with 2 poggles in each row

Refer to the **Collaborate** task of the **Pupil Resource**.

Ask children to compare their chosen arrangement with others.

- *Compare your ideal arrangement with someone else. What is different? What is the same?*
- *Do you agree with each other? Who gives the most convincing reasons?*

Children could then work in pairs in the **Communicate** task to produce their brochure, ideally partnering up with someone else who has chosen the same arrangement.

Refer to the **Communicate** task of the **Pupil Resource**.

Slide 13

Display slide 13.

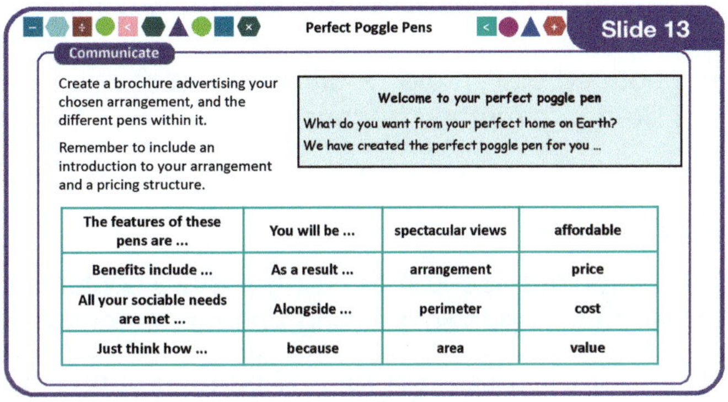

Ask children to take on the role of an estate agent or a developer and create a brochure advertising their chosen arrangement, and the different pens available within it.

Children could create a pricing structure. Here are some things that could influence the pricing of each pen within an arrangement.

- *As poggles are sociable, would a plot where they are surrounded by neighbours be more desirable? How much would a middle pen cost?*
- *Would a plot on the outside of an arrangement, and so with nice views of the farm, be higher priced? How much would you charge for an outer pen?*

Encourage children to use the word bank and the writing frame to help them with their persuasive writing.

The features of these pens are ...	**You will be ...**	**spectacular views**	**affordable**
Benefits include ...	**As a result ...**	**arrangement**	**price**
All your sociable needs are met ...	**Alongside ...**	**perimeter**	**cost**
Just think how ...	**because**	**area**	**value**

Refer to the **Create** task of the **Pupil Resource**.

- *Were there any questions asked at the start of the task that you now want to investigate?*
- *Is there anything new that you are wondering now?*

Children can use the word bank to prompt further questions of their own.

2 metres by 2 metres	**unequal groups**	**arrangements**

- I wonder if I had fewer poggles, would I be able to make fewer arrangements.
- I wonder what would happen if the rows could have unequal groups. Could I still make the same number of arrangements?
- I wonder if I could make arrangements where the sides don't have to meet. How many could I make?
- I wonder what would happen if each poggle needed 2 metres by 2 metres.

Consider

Refer to the **Consider** task of the **Pupil Resource**.

Slide 14

Display slide 14.

Encourage children to think about the strategies that they have used. The prompts below may help get children started.

Cognitive learning domain

- I am sure / unsure of my answer because …
- The best strategy was …
- I could include more proofs by ...

Affective domain

- The fun bit of my learning was … It was fun because …
- I noticed that … did … differently from me.
- I liked …'s method because ...

Metacognitive domain

- I got stuck when … I unstuck myself by …
- I used … to help me. It was helpful because …
- Next time I try to solve a problem, I will …

Answers

If we take into account that poggles are sociable creatures and want to be close to each other, there are nine possible arrangements that ensure that all pens are next to at least two other pens.

Number of rows	Number of pens in each row	Perimeter (metres)	Area (metres squared)
1	36	74 m	36 m^2
2	18	40 m	36 m^2
3	12	30 m	36 m^2
4	9	26 m	36 m^2
6	6	24 m	36 m^2
9	4	26 m	36 m^2
12	3	30 m	36 m^2
18	2	40 m	36 m^2
36	1	74 m	36 m^2

The area of any arrangement is always 36m^2.

The perimeter will change depending on the arrangement. 'Pairs' of arrangements will have equal perimeters. For example, the perimeter of 4 rows of 9 pens will be the same as 9 rows of 4 pens.

What if the number of poggles was different?

Some children may assume that the smaller the number of poggles, the fewer arrangements are possible. They may be surprised to find that, for example, there are six possible arrangements for 18 poggles, but only two for 19 poggles.

What if each poggle pen had to be 2 metres by 2 metres?

Some children may assume that, if each pen increases from 1 m by 1 m to 2 m by 2 m, the area of the whole arrangement will double. Children should be able to make links within their multiplication knowledge (building models using cubes or square sticky notes as needed) to work out that if each poggle pen is 2 metres by 2 metres, this is the same as four 1 metre by 1 metre squares placed together, and therefore each pen has an area of 4 metres squared. Therefore the area of the whole arrangement will be 4 times as large ($4 \times 36\ m^2 = 144\ m^2$).

Shapes Within Shapes

Year 3

Introduction

Shapes Within Shapes provides an opportunity for children to apply their knowledge of shapes. During the **Curious** task, children will consider what shapes they can see within the given square and triangle. It is important to recognise shapes in different orientations and compare them. This task is rich in opportunities to develop different lines of enquiry, with children noticing and communicating connections and patterns.

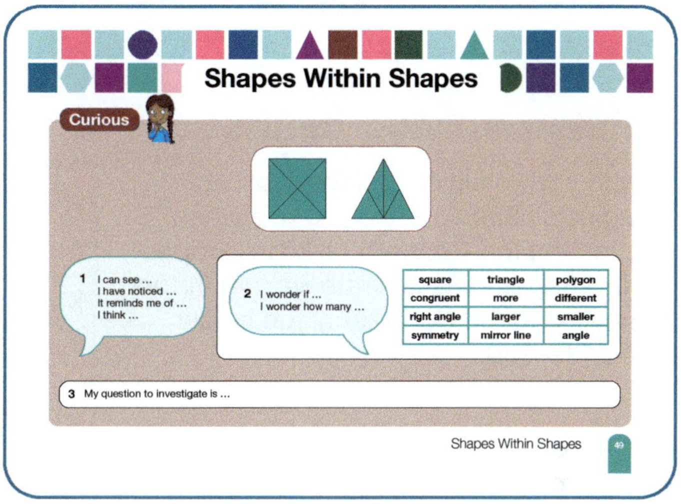

Throughout the task, children should be encouraged to confidently use precise vocabulary to describe the shapes created and their properties. Teachers can choose whether to introduce further vocabulary, such as 'congruent' and specific properties of quadrilaterals, as appropriate. Following their exploration, children should compare and contrast what they have found out and identify multiple learning outcomes.

This task promotes greater-depth learning in a range of ways, supporting children to identify and articulate underlying mathematical structures. For example, in the **Consider** task, greater depth is promoted by asking children to identify learning behaviours and strategies that were and were not useful, and to reflect on what was and was not engaging about the task.

This **Teacher Guide** is supported by photocopiable **Pupil Resources** and **Teacher's Presentation** PowerPoint slides for whole-class or group display.

National Curriculum

Shapes Within Shapes covers these Year 3 National Curriculum objectives.

Geometry – properties of shapes

- Draw 2D shapes and make 3D shapes using modelling materials; recognise 3D shapes in different orientations and describe them.
- Recognise angles as a property of shape or a description of a turn.
- Identify right angles, recognise that two right angles make a half-turn, three make three-quarters of a turn and four a complete turn; identify whether angles are greater than or less than a right angle.
- Identify horizontal and vertical lines and pairs of perpendicular and parallel lines.

The 6C cycle

The grid on the right shows the relative weighting given to each of the 6 Cs in **Shapes Within Shapes**.

Curious	★★★
Connect	★★
Collaborate	★★★
Communicate	★★
Create	★★
Consider	★★★

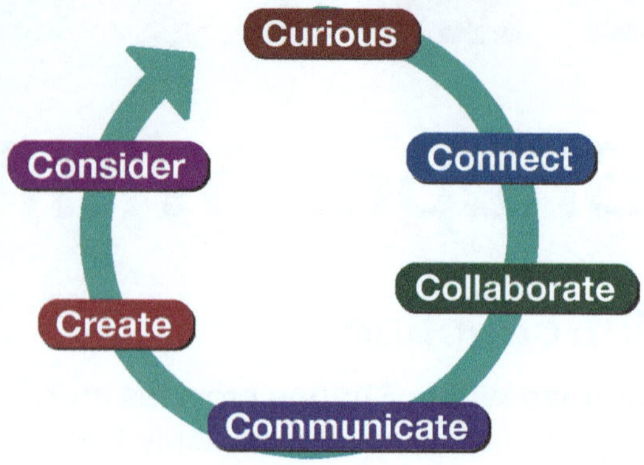

Problem-solving heuristics

The grid shows which problem-solving strategies children are likely to use when investigating the Connect tasks in **Shapes Within Shapes**. These strategies are further exemplified in the Connect stage below (pages 139–140).

Draw a diagram	✓
Make a table / spot a pattern	✓
Make a model / act it out	✓
Try then improve	
Create a systematic list	
Solve a simpler but related problem	
Work backwards	

Key mathematical ideas

Shapes Within Shapes provides an opportunity for children to apply their knowledge of shapes, including their properties, and rehearse specific vocabulary. They will explore the shapes that they can create by dividing larger shapes using straight lines.

Children should be encouraged to use the following vocabulary, as well as accurate shape names, when discussing and describing shapes: sides, vertex (plural: vertices), polygon (2D shape with straight sides), curved side, regular, irregular, symmetrical, non-symmetrical, angle, right angle, smaller than a right angle, greater than a right angle, internal angle, horizontal, vertical, perpendicular, parallel, diagonal, congruent (identical properties including length of sides and angles).

Children may need reminders of the properties of some of the shapes.

- Right-angled triangle: one angle is a right angle (90 degrees).
- Isosceles triangle: two sides are equal in length.
- Scalene triangle: no sides are equal in length.
- Rectangle – four straight sides and four right angles. A square is a special type of rectangle with all four sides equal in length.
- Rhombus – four straight sides all of equal length and two pairs of equal angles. A square is a special type of rhombus.
- Parallelogram – two pairs of straight opposite sides that are parallel to each other. A rhombus is a special type of parallelogram with all four sides equal in length.
- Kite – four straight sides with adjacent pairs of sides having an equal length.
- Trapezium – four straight sides with one pair of parallel lines.

Answers to the tasks are given on pages 142–143.

Teaching guidance

If you are using the **Express Pathway**, please turn to pages 138–142.

Teaching guidance for the **Full 6C Pathway** is given below.

The **Teacher's Presentation** PowerPoint file has slides specifically mentioned in this teaching guidance. Some of the slides include notes (within the PowerPoint file) providing 'stumble support' suggestions, that may help you to support children in their problem-solving thinking and enable access for children who are less confident.

Awe and wonder

Slide 1

Display slide 1.

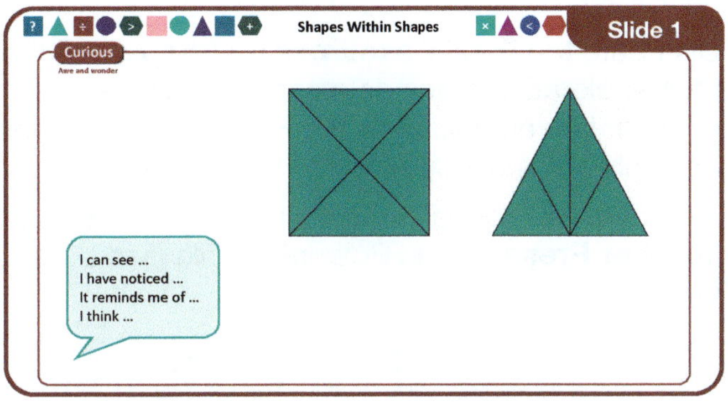

- I can see a square and a triangle.
- I have noticed that there are extra lines through each shape.
- I can see 8 triangles inside the two shapes.
- I have noticed that the four triangles inside the square all contain a right angle, but the triangles within the larger triangle do not.
- I think that the triangles within the square are congruent / the same, but I don't think triangles within the larger triangle are.
- It reminds me of when we learned about symmetry. The line in the middle is like a mirror line.

⚠ If children are not familiar with the meaning of the word 'congruent', it would be useful to introduce / reinforce this vocabulary now.

Develop and deepen

Slide 2

Display slide 2.

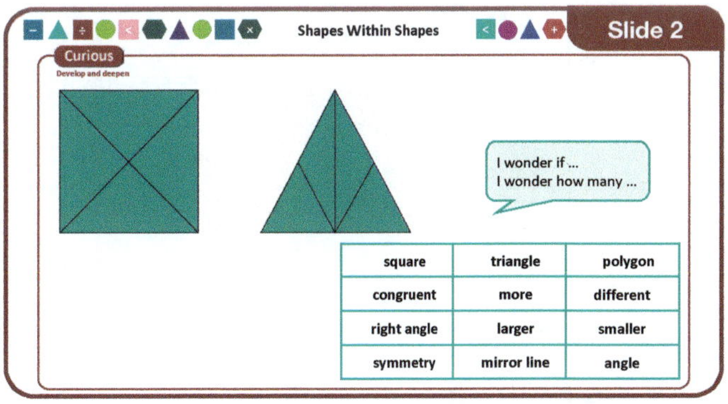

Encourage children to use the words in the word bank to develop and deepen their thoughts and questions.

square	**triangle**	**polygon**
congruent	**more**	**different**
right angle	**larger**	**smaller**
symmetry	**mirror line**	**angle**

Possible lines of enquiry linked to the word bank:

- I wonder how many more triangles I can make within these shapes.
- I wonder if I could make the new triangles in the larger triangle all the same.
- I wonder if I can draw right-angled triangles inside the larger triangle.
- I wonder if all polygons can be split into triangles using just one line.
- I wonder if I could make different smaller shapes inside big shapes (for example, squares).

Possible lines of enquiry that children may raise now, but should be held back until the **Create** task:

- I wonder if the smaller triangles inside the large triangle have the same area / take up the same amount of space.
- I wonder how many new shapes / polygons I could make if I cut up a square into triangles and rearrange the triangles.
- I wonder if I could make triangles from the inside of a circle.

Consider whether / how to collate and display all the suggested questions, and whether / how to steer children towards selecting an appropriately challenging question.

Pupil resources for Full and Express Pathways

Distribute the **Pupil Resources** for **Shapes Within Shapes** to the children to help support them in their exploration of the different tasks. Use either of the **Curious** **Pupil Resources** depending on whether you are following the **Full Pathway** or the **Express Pathway**. The **Pupil Resources** for the remaining 5Cs are common to both pathways.

To provide additional flexibility in the classroom, each of the **Pupil Resources** is also provided in the **Teacher Presentation** PowerPoint file (see slides 3–6), so you can display these at any time if you wish.

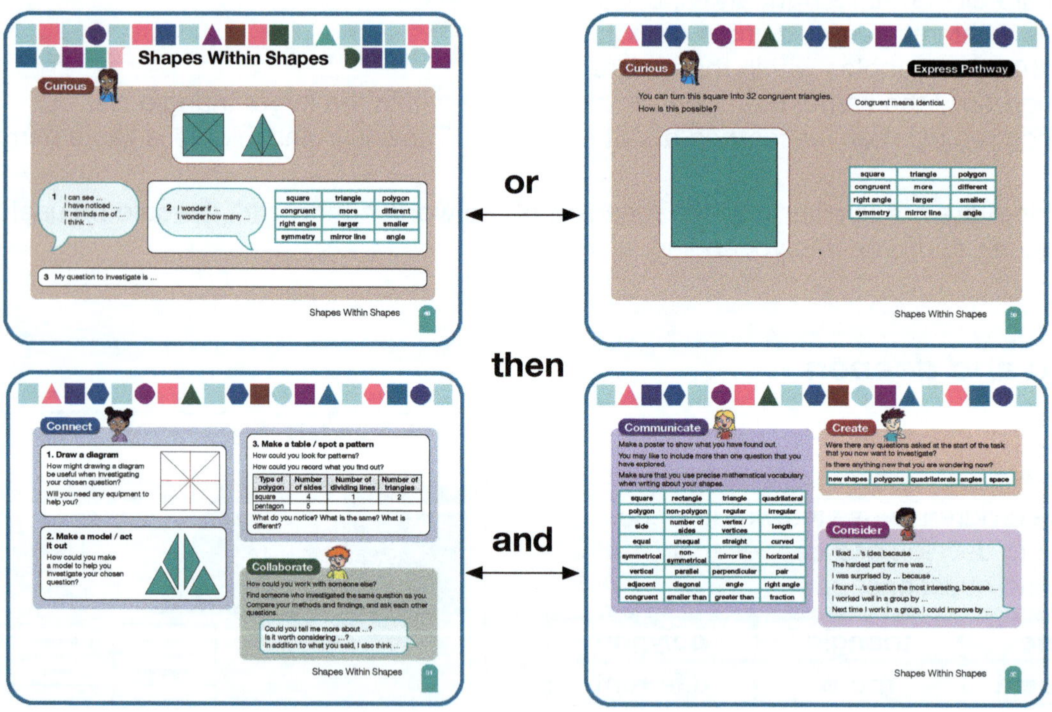

Express Pathway

If you are following the Express Pathway, spend a few minutes discussing the **Curious Pupil Resource** with the children. (This is slide 4, so you could display it if you wish.) Use some of the prompts on the previous page to help.

Three tasks involving different strategies are suggested for children to make connections.

1. Draw a diagram

Additional resources required: rulers, set squares (or homemade right-angle measurers)

Slide 7

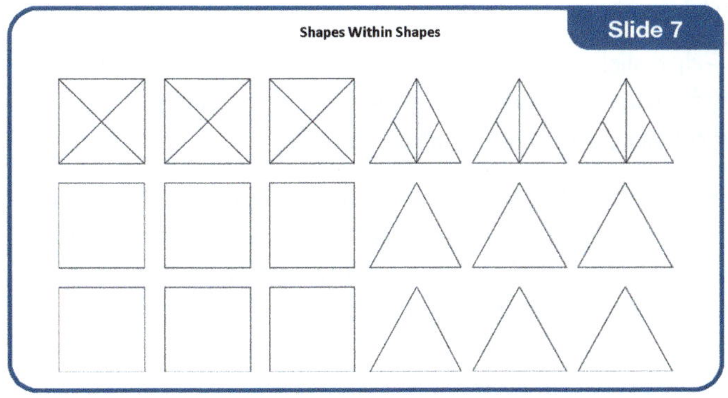

Print slide 7 and distribute to the children. They can use this to draw new or additional lines to divide the square and triangle, to test out their ideas.

- *How might drawing a diagram be useful when investigating your chosen question?*
- *Will you need any equipment to help you?*

This method could be used to explore several of the lines of enquiry. For example, 'I wonder if I could draw right-angled triangles inside the larger triangle.'

- *Do all of the smaller triangles have a right angle at the moment?*
- *How could you check this?*
- *Could you draw any additional lines to make all the shapes inside the triangle right-angled triangles?*

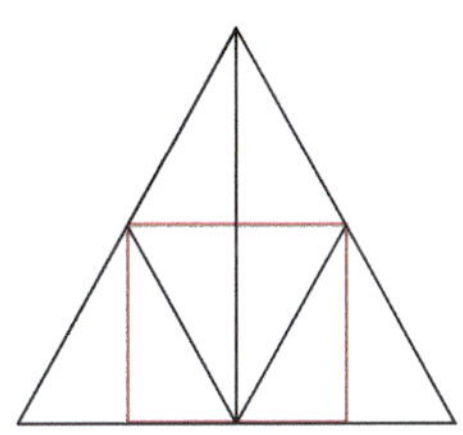

2. Make a model / act it out

Additional resources required: scissors, geoboards (optional), digital geoboards (optional)

Slide 8

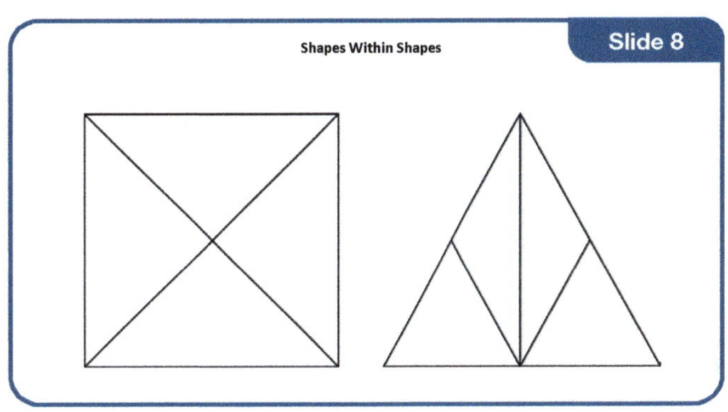

Print slide 8 (ideally on card) and distribute to the children.

Encourage children to accurately cut out the smaller shapes, and use them to build models of the shapes.

For example, if investigating the line of enquiry: 'I wonder if I could make the new triangles in the larger triangle all the same', children could lay the parts of the triangle on top of each other as a way of comparing the size, then make additional cuts and compare again.

Or they could explore how many more triangles they can make within the shapes, by splitting the shape-parts into smaller triangles.

Children could also experiment with using the cut-out parts to see if they can be put together in different ways to make any other shapes.

Geoboards or digital geoboards would also be useful for making models.

3. Make a table / spot a pattern

Additional resources required: rulers

This strategy would be useful if children are exploring the line of enquiry: 'I wonder if all polygons can be split into triangles using just one line.'

Slide 9

Print slide 9 and distribute to the children. They can use this to draw lines within the shapes to see how triangles can be made.

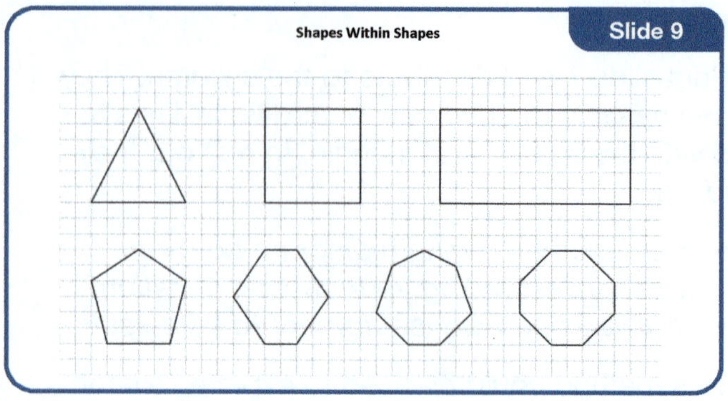

- I can split the triangle, the square and the rectangle into triangles by drawing a single line. But it doesn't work for the pentagon, hexagon, heptagon or octagon.

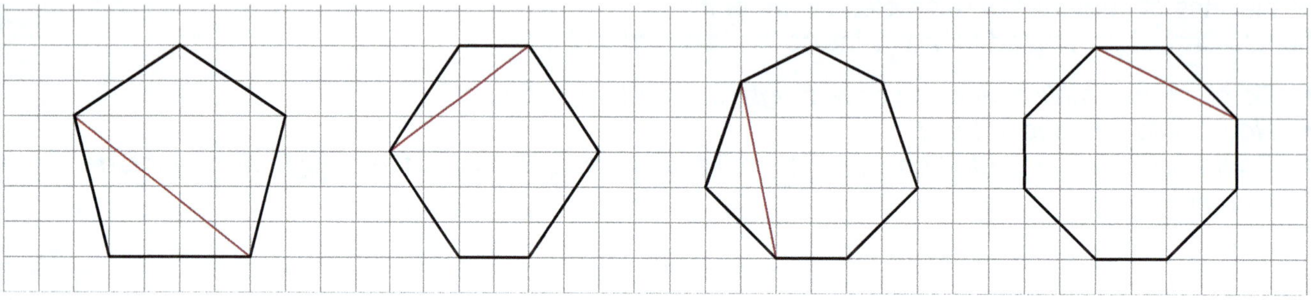

- For these shapes, I need to draw more lines.

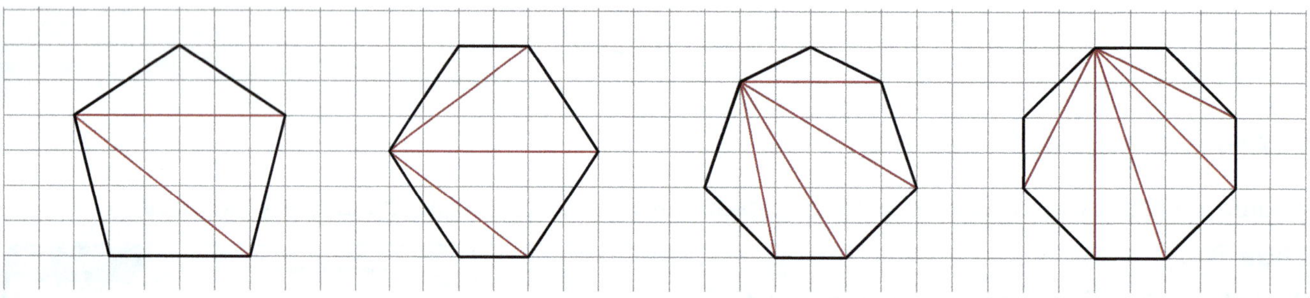

Encourage children to put this information into a table.

Slide 10

Display slide 10, if needed.

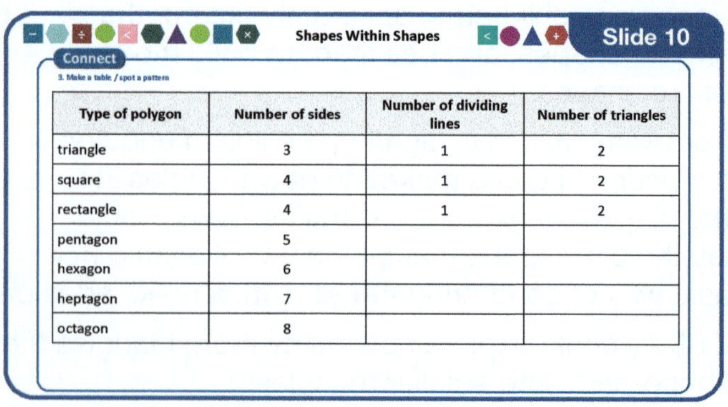

Type of polygon	Number of sides	Number of dividing lines	Number of triangles
triangle	3	1	2
square	4	1	2
rectangle	4	1	2
pentagon	5		
hexagon	6		
heptagon	7		
octagon	8		

- *How many lines are needed to split each type of polygon into triangles?*
- *How many triangles does this give you in each polygon?*
- *Do you notice anything about the number of sides and the number of triangles that you have made?*
- The number of triangles in each shape is 2 less than the number of sides.
- I think that a nonagon will have seven triangles because 9 – 2 is 7.
- *What happens if the lines are allowed to cross? How many triangles would there be in each polygon?*

Refer to the **Collaborate** task of the **Pupil Resource**.

Slide 11

Display slide 11.

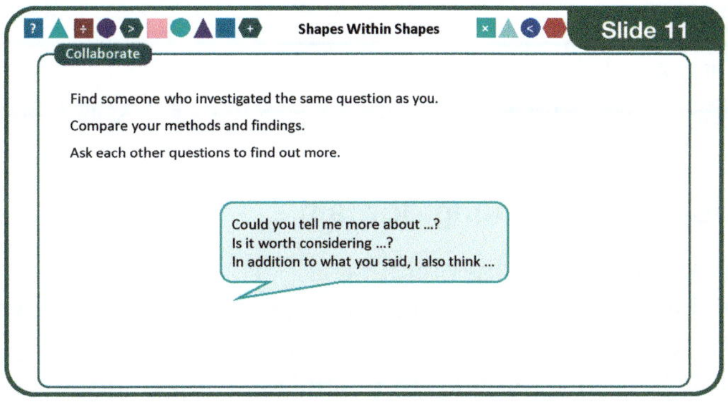

Ask children to pair up with a child who investigated the same line of enquiry. They could start by comparing their methods and findings. The following sentence stems would be useful to encourage children to question and build upon each other's ideas:

- Could you tell me more about …?
- Is it worth considering …?
- In addition to what you said, I also think …

Children could then work in pairs or groups to create posters of their lines of enquiry, as detailed below in the **Communicate** task.

Refer to the **Communicate** task of the **Pupil Resource**.

Slide 12

Display slide 12.

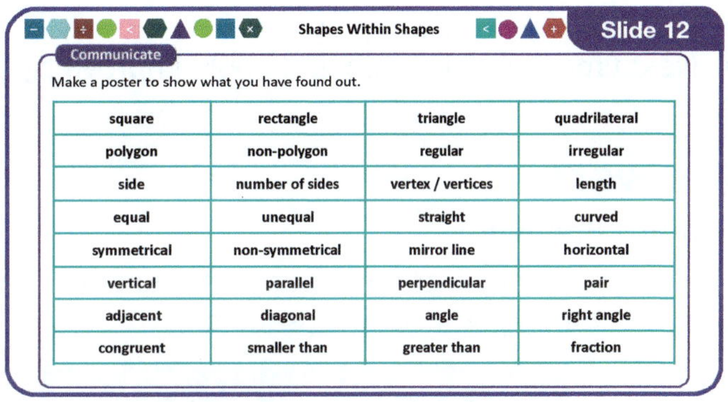

square	rectangle	triangle	quadrilateral
polygon	non-polygon	regular	irregular
side	number of sides	vertex / vertices	length
equal	unequal	straight	curved
symmetrical	non-symmetrical	mirror line	horizontal
vertical	parallel	perpendicular	pair
adjacent	diagonal	angle	right angle
congruent	smaller than	greater than	fraction

This task would provide a good opportunity for children to create a poster, to illustrate and explain what they found out when exploring one of their lines of enquiry.

Encourage children to use the word bank to support them in their use and spelling of precise mathematical language.

Children could also be tasked to consider and explain verbally what they think was the most interesting line of enquiry they followed and why. These questions could be used to support children's discussions:

- *What did you most enjoy about this investigation?*
- *What were you surprised about? Why?*
- *Did anyone else's question interest you? Why?*

Refer to the **Create** task of the **Pupil Resource**.

Slide 13

Display slide 13.

- *Were there any questions asked at the start of the task that you now want to investigate?*
- *Is there anything new that you are wondering now?*

Children can use the word bank to prompt further questions of their own.

new shapes	**polygons**	**quadrilaterals**	**angles**	**space**

- I wonder if I could make triangles from the inside of a circle.
- I wonder if the smaller triangles in the large triangle have the same area / take up the same amount of space.
- I wonder how many new shapes / polygons I could make if I cut up a square into triangles and rearrange the triangles.

Consider

Refer to the Consider task of the **Pupil Resource.**

Slide 14

Display slide 14.

Encourage children to think about the strategies that they have used. The prompts below may help get children started.

Cognitive learning domain

- I liked …'s idea, because …
- I learned … from someone else in the classroom.
- The hardest part for me was …
- If I was to investigate a similar problem again, I have learned ...

Affective domain

- I enjoyed …
- I was surprised by … because …
- I found ....'s question the most interesting, because ...

Metacognitive domain

- I worked well in a group by …
- Next time I work in a group, I could improve by ...

Answers

Express Pathway

These are examples of the processes and patterns that children might follow to create 32 congruent triangles inside the square.

Children should spot a link between the patterns that they can see developing and fractions.

- If I draw two lines, the square is divided into quarters. If I draw another two lines, I get eighths. 8 is double 4, but eighths are half the size of quarters.

I wonder how many more triangles I can make within these shapes.

In theory, children could keep dividing the triangles further and further to create more triangles, with no end point.

I wonder if all the triangles in the large triangle could be the same / I wonder if I could draw right-angled triangles inside the larger triangle.

By dividing each of the existing triangles in half, six congruent, right-angled triangles can be created.

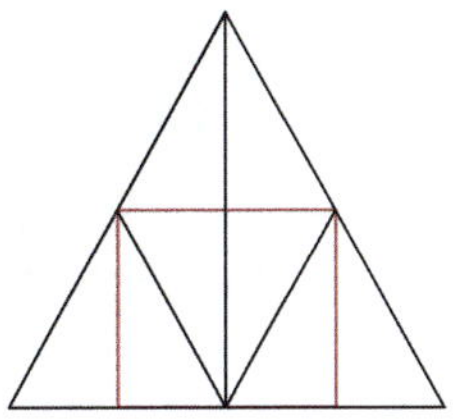

I wonder if all polygons can be split into triangles using just one line.

Only triangles and quadrilaterals can be divided into triangles using just one.

If children explore how many lines are needed to divide other polygons into triangles, they should discover that:

- the number of dividing lines is 3 less than the number of sides;
- the number of triangles produced is 2 less than the number of sides;
- the number of triangles produced is 1 more than the number of dividing lines.

Type of polygon	Number of sides	Number of dividing lines	Number of triangles
square / rectangle	4	1	2
pentagon	5	2	3
hexagon	6	3	4
heptagon	7	4	5
octagon	8	5	6

If children then extend this to explore what happens if all diagonals of the polygons are joined, children should discover the following:

Type of polygon	Number of sides	Number of dividing lines
square / rectangle	4	2
pentagon	5	5
hexagon	6	9
heptagon	7	14
octagon	8	20

Children may not be able to offer an explanation for what they find, but the number of dividing lines can be found by calculating: [number of vertices] × [number of vertices – 3] ÷ 2.

L Plates

Year 4

Introduction

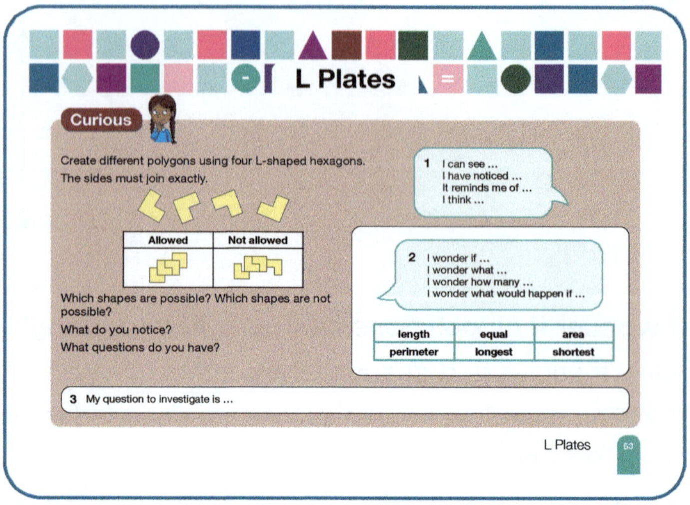

L Plates allows children to think deeply about perimeter and use a range of heuristics to find, and prove they have found, the minimum and maximum perimeters possible. Children explore which polygons are possible to make using the four hexagonal 'L plates' and start to notice how the perimeters can vary widely.

Children who choose to make models or draw diagrams can observe how the four hexagons interact and influence the perimeter of the polygon created. Some may use a 'try then improve' strategy and develop efficiency by tweaking their arrangement rather than starting from the beginning every time. Children can also make a table to support systematic thinking. This task supports children to make conjectures about a simpler problem involving two hexagons, then apply these to four hexagons.

This task promotes greater–depth learning in a range of ways, supporting children to identify and articulate underlying mathematical structures. For example, in the **Curious** stage, greater depth is promoted by asking children to make rich mathematical contributions to discussions and suggest a variety of appropriate mathematical questions.

This **Teacher Guide** is supported by photocopiable **Pupil Resources** and **Teacher's Presentation** PowerPoint slides for whole–class or group display.

National Curriculum

L Plates covers these Year 4 National Curriculum objectives.

Geometry – properties of shapes

- Compare and classify geometric shapes, including quadrilaterals and triangles, based on their properties and sizes.

Measurement

- Measure and calculate the perimeter of a rectilinear figure (including squares) in centimetres and metres.

The 6C cycle

The grid on the right shows the relative weighting given to each of the 6 Cs in **L Plates**.

Curious	★★★
Connect	★★
Collaborate	★
Communicate	★★
Create	★★
Consider	★★

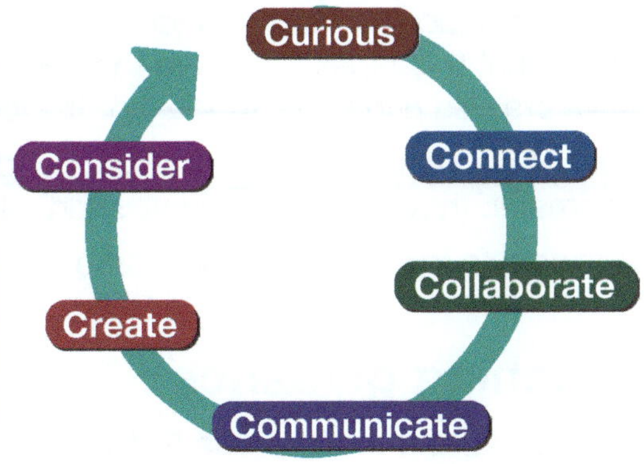

Problem–solving heuristics

The grid shows which problem–solving strategies children are likely to use when investigating the **Connect** tasks in **L Plates**. These strategies are further exemplified in the **Connect** stage below (pages 148–149).

Draw a diagram	✓
Make a table / spot a pattern	✓
Make a model / act it out	✓
Try then improve	✓
Create a systematic list	
Solve a simpler but related problem	✓
Work backwards	

Key mathematical ideas

Start by discussing the L–shapes, and how we know they are hexagons. This could lead to a conversation about regular and irregular polygons. When children are counting the sides of the L–shapes, ensure they mark where they started counting.

For the purposes of this investigation, here are the dimensions of the L–shape. The rules of the investigation state that sides must join exactly to ensure that the perimeters of all the shapes created remain in whole numbers.

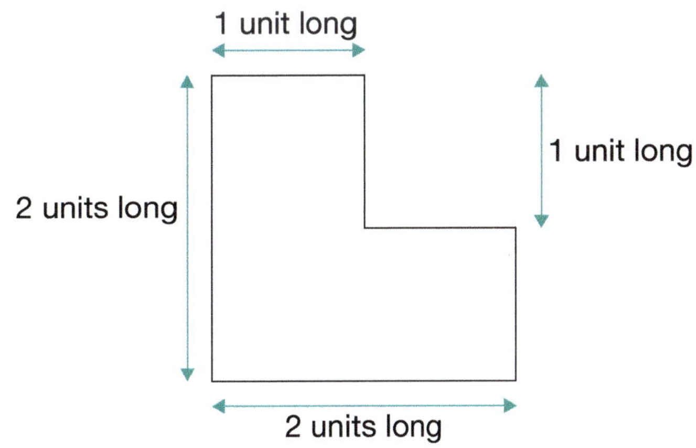

If needed, explain the term 'congruent': congruent shapes have identical properties including length of sides and angles. Some children may not recognise that two shapes are congruent if they have different orientations.

When children are exploring the different shapes they can make, ensure they are counting the sides correctly. Ask them to name the shapes they are creating and some of the properties. For example, are they symmetrical, are they regular polygons? Also, discuss that when a shape is made up of two or more geometric shapes it can be referred to as either a composite or a compound shape.

As children explore the perimeters of their composite shapes, ensure that all children understand what the perimeter is and how to calculate it. Can they come up with different methods for calculating the perimeter? Look out for children who may not realise that when two shapes are joined, the two lengths that are connected are excluded from the perimeter of the composite shape.

Then introduce the task of finding the shortest and the longest perimeter when joining the four congruent L–shapes. What are the perimeters of these shapes? How can children prove that they have created either the shortest or the longest perimeter?

For the shortest perimeter, they will need as many sides touching as possible. For the longest perimeter, they will need as few, short, sides touching as possible to increase the total perimeter.

Answers to the tasks are given on page 151.

Teaching guidance

If you are using the **Express Pathway**, please turn to pages 148–151.

Teaching guidance for the **Full 6C Pathway** is given below.

The **Teacher's Presentation** PowerPoint file has slides specifically mentioned in this teaching guidance. Some of the slides include notes (within the PowerPoint file) providing 'stumble support' suggestions, that may help you to support children in their problem–solving thinking and enable access for children who are less confident.

Awe and wonder

Slide 1

Print slide 1 and distribute to the children. They can cut out the four shapes and manipulate them to see what they observe. (An alternative would be to ask children to build the L–shapes using interlocking cubes.)

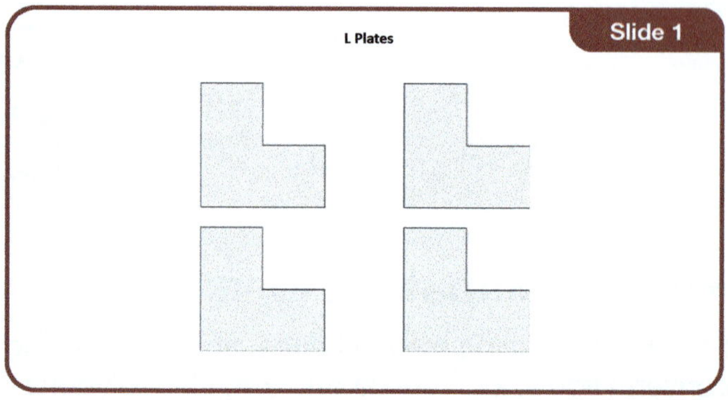

Slide 2

Display slide 2.

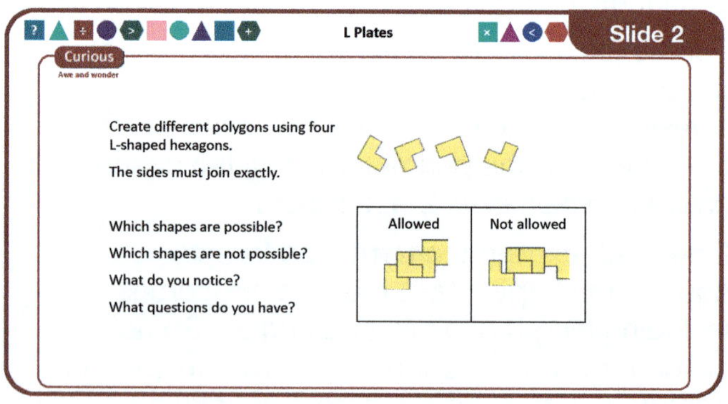

- <u>I can see</u> that they are exactly the same size and shape.
- <u>I have noticed</u> they can fit together to make other shapes.
- <u>I can see</u> that they are 'L–shapes' and some are upside down.
- <u>I can see</u> that the two longer sides are twice as long as the four shorter sides.
- <u>I have noticed</u> that the 'L–shape' is a hexagon because it has six sides.
- <u>It reminds me of</u> when we did work on symmetry.

⚠ Ensure that all children know and understand the meaning of the words 'congruent' and 'composite'. Check that children recognise that two shapes are congruent even if they have different orientations.

Develop and deepen

Encourage children to use the words in the word bank to develop and deepen their thoughts and questions.

length	**equal**	**area**
perimeter	**longest**	**shortest**

Possible lines of enquiry:

- I wonder which shapes are possible to create using the four hexagons.
- I wonder if you can make a square / rectangle / hexagon.
- I wonder what is the most / least amount of sides.

Possible lines of enquiry linked to the word bank:

- I wonder what is the longest / shortest perimeter possible.

Possible lines of enquiry that children may raise now, but should be held back until the **Create** task:

- I wonder what would happen if I used another base shape.

Consider whether / how to collate and display all the suggested questions, and whether / how to steer children towards selecting an appropriately challenging question.

Pupil Resources for Full and Express Pathways

Distribute the **Pupil Resources** for **L Plates** to the children to help support them in their exploration of the different tasks. Use either of the **Curious Pupil Resources** depending on whether you are following the **Full Pathway** or the **Express Pathway**. The **Pupil Resources** for the remaining 5Cs are common to both pathways.

To provide additional flexibility in the classroom, each of the **Pupil Resources** is also provided in the **Teacher's Presentation** PowerPoint file (see slides 3–6), so you can display these at any time if you wish.

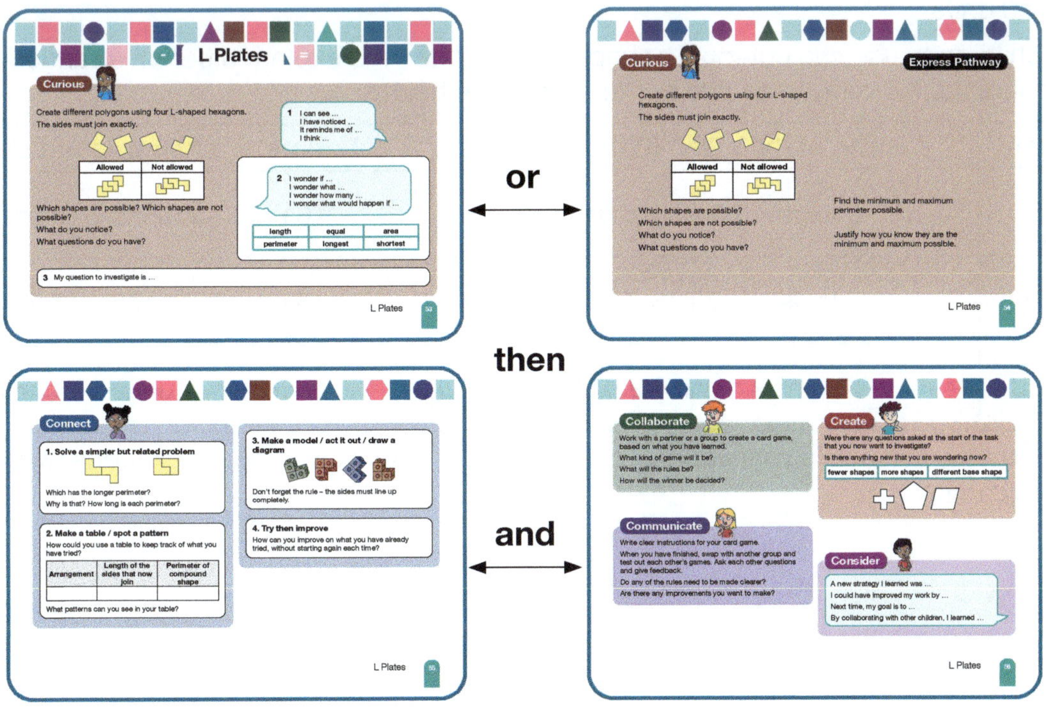

Express Pathway

If you are following the **Express Pathway**, spend a few minutes discussing the **Curious Pupil Resource** with the children. (This is slide 4, so you could display it if you wish.) Use some of the prompts on the previous page to help.

Four tasks involving different strategies are suggested for children to make connections.

1. Solve a simpler but related problem

Slide 7

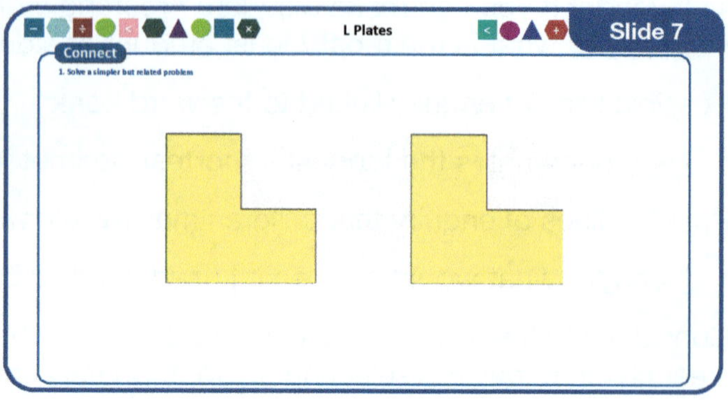

Display slide 7.

Explain that, throughout the whole task, children will not be measuring the perimeters of their shapes; instead, they will be calculating them as a number of 'units'. Each of the long sides of a hexagon is 2 units. Each of the short sides of a hexagon is 1 unit.

Children could start by working with just two L–shaped hexagons.

- Each L–shaped hexagon has a perimeter 8 units long.
- The total perimeter of these two hexagons, when they are not joined together, is 16 units (8 units multiplied by 2).

To maximise the perimeter, the length of where the hexagons join must be minimised, meaning children need to have as little of the two shapes touching as possible. In this scenario, the minimum join is to have two sides of length 1 unit touching. This will reduce the total perimeter by 2 units.

To minimise the perimeter, the length of where the hexagons join must be maximised. The maximum length of join for two hexagons is the two shapes connecting along three sides, reducing the total perimeter by 6 units.

Encourage children to use two cut–out L–shapes (from slide 1) to explore different arrangements, and record their findings, possibly in a table similar to the one below.

Arrangement	Length of the sides that now join	Perimeter of compound shape
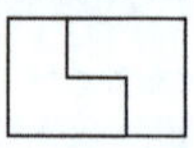	2 joins × 3 units = 6 units	16 units – 6 units = 10 units
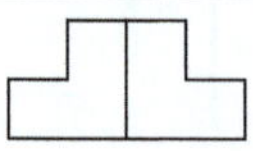	2 joins × 2 units = 4 units	16 units – 4 units = 12 units
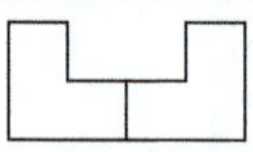	2 joins × 1 unit = 2 units	16 units – 2 units = 14 units

Children could then use this logical approach to solve the task involving four hexagons, with a total perimeter of 32 units.

2. Make a table / spot a pattern

Slide 8

Print slide 8 and distribute to children to use to keep track of the models they are building or drawing.

Arrangement	Length of the sides that now join	Perimeter of compound shape
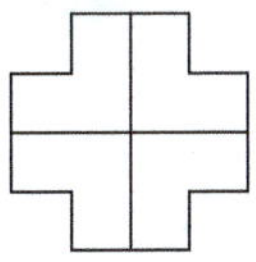	8 joins × 2 units = 16 units	32 units – 16 units = 16 units
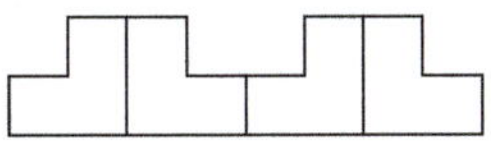	4 joins × 2 units = 8 units 2 joins × 1 unit = 2 units	32 units – 10 units = 22 units

Ensure children remember that for each join, they are losing two sides, one from each hexagon.

⚠ Look out for children not joining the sides of the L–shapes in full. This rule must be followed so the perimeters remain in whole numbers.

3. Make a model / act it out / draw a diagram

Additional resources required: interlocking cubes, squared paper, dotted paper

Slide 9

Display slide 9.

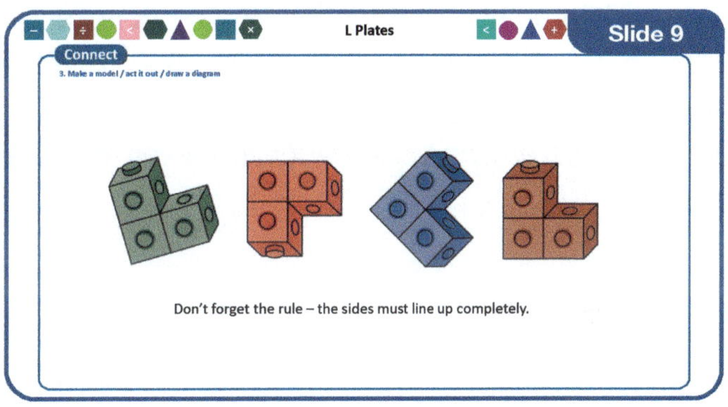

Children may choose to physically manipulate the L–shapes (cut out from printouts of slide 1). Alternatively, they could build the L–shapes using interlocking cubes; they could use three cubes in a different colour for each L–shape.

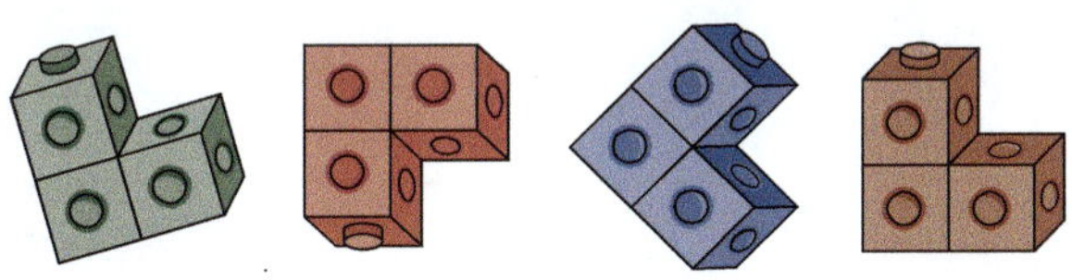

Some children may prefer to draw diagrams using squared paper or dotted paper.

4. Try then improve

Children may choose a 'trial and improvement' approach to find the shortest or longest perimeter possible. If this is the case, encourage children to build on what they have learned, rather than recreating their model / diagram from scratch each time they try again.

Collaborate

Refer to the **Collaborate** task of the **Pupil Resource**.

Ask children to work in pairs or groups to create a card game based on what they have learned about how the arrangement of a set of identical shapes affects the perimeter.

Encourage children to discuss the things they need to think about when planning a game.

- *What kind of game will it be?* (barrier game / match the rule card game / sorting to a rule game)
- *What will the rules be?*
- *What base shapes will your game use? How many shapes must be used?*
- *How will the winner be decided?*

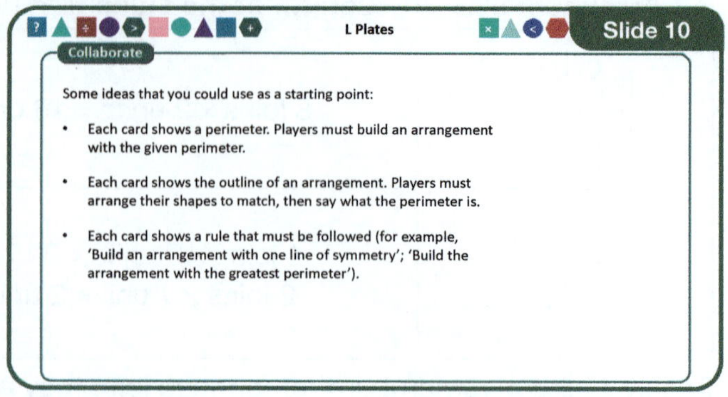

Slide 10

Display slide 10 if needed. This shows some ideas for games that children could use as a starting point if they are struggling to come up with a concept.

Communicate

Refer to the **Communicate** task of the **Pupil Resource**.

Ask children to write clear instructions for the game they invented in the **Collaborate** task.

- *Remember, people will need to know: how many players can participate; what resources will they need; what the rules are; how the winner is decided.*

When children have written their instructions, ask them to swap with another group and test out each other's games. They can then ask each other questions and give feedback.

- *Do any of the rules need to be made clearer?*
- *Are there any improvements you want to make?*

Create

Refer to the **Create** task of the **Pupil Resource**.

Slide 11

Display slide 11.

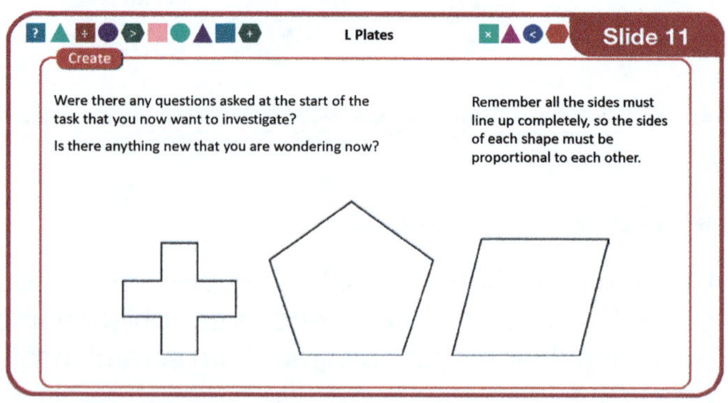

- *Were there any questions asked at the start of the task that you now want to investigate?*
- *Is there anything new that you are wondering now?*

Children can use the word bank to prompt further questions of their own.

fewer shapes	**more shapes**	**different base shape**

- I wonder what would happen if I used another base shape.
- I wonder what would happen if I used more or fewer 'L–shape' hexagons.

Children may choose to investigate other base shapes such as these: 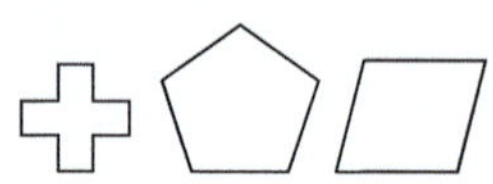

Refer to the Consider task of the **Pupil Resource**.

Slide 12

Display slide 12.

Encourage children to think about the strategies that they have used. The prompts below may help get children started.

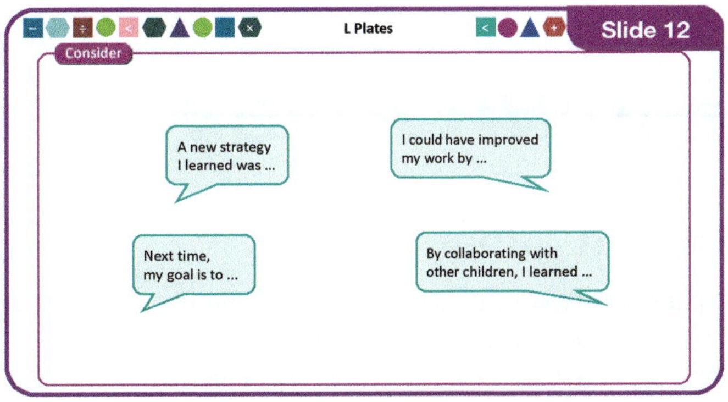

Cognitive learning domain

- I learned …
- I could use this information in the future when ...
- A new strategy I learned was …

Affective domain

- I am pleased with my answer because …
- I could have improved my work by …
- Next time, my goal is to ...

Metacognitive domain

- I learned this about working in a group: …
- By collaborating with other children, I learned …
- I could become a better group member by ...
- … helped me by …

Answers

The perimeter of each L–shaped hexagon is 8 units.

The total perimeter of all four L–shaped hexagons, when not joined, is 32 units.

When joined together, the shortest possible perimeter is 14 units.

32 units – 18 units = 14 units

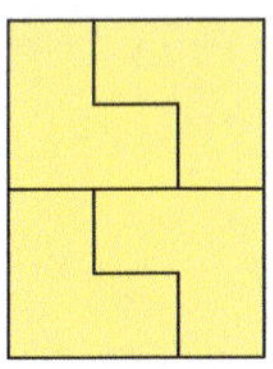

The longest possible perimeter is 26 units.

32 units – 6 units = 26 units

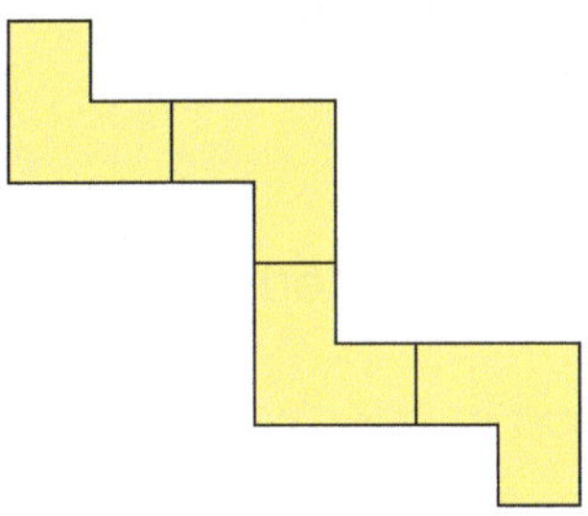

Blown Away

Year 3
Year 4

Introduction

In **Blown Away**, children will assume the role of hovercraft engineers and attempt to persuade others that their design is best. They first explore what 'the best' really means, deepening learning by understanding that this is not a fixed idea and more than one answer is possible. Once they have established how to improve their hovercraft, they identify which variable to change and how to measure and record the impact of their modification.

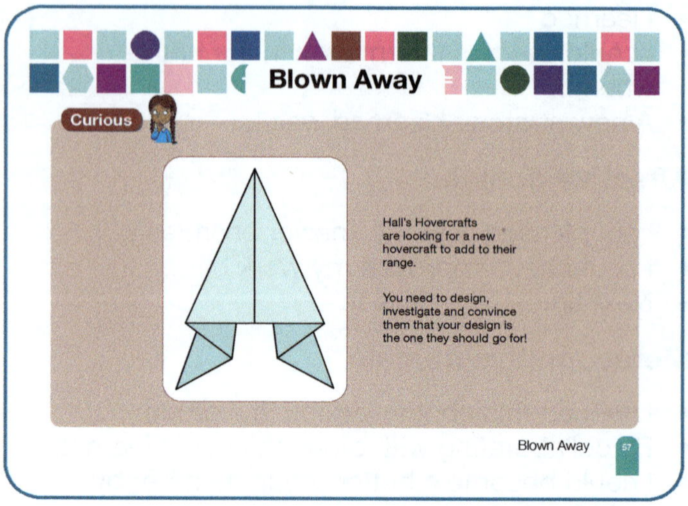

The task involves children working collaboratively, establishing group roles and developing effective team-working strategies including listening and building on the ideas of others. Children must then communicate their findings, interpreting their data into a clear and persuasive argument, using accurate mathematical language and diagrams.

This task promotes greater-depth learning in a range of ways, supporting children to identify and articulate underlying mathematical structures. For example in the **Collaborate** stage, greater depth is promoted by asking children to offer suggestions, listen carefully to others and build upon their own ideas, and be able to accept it if their ideas are rejected by the group.

This **Teacher Guide** is supported by photocopiable **Pupil Resources** and **Teacher's Presentation** PowerPoint slides for whole-class or group display.

National Curriculum

Blown Away covers these Year 3 and Year 4 National Curriculum objectives.

Year 3 Statistics

- Interpret and present data using bar charts, pictograms and tables.
- Solve one-step and two-step questions [for example, 'How many more?' and 'How many fewer?'] using information presented in scaled bar charts and pictograms and tables.

Year 4 Statistics

- Interpret and present discrete and continuous data using appropriate graphical methods, including bar charts and time graphs.
- Solve comparison, sum and difference problems using information presented in bar charts, pictograms, tables and other graphs.

The 6C cycle

The grid to the right shows the relative weighting given to each of the 6 Cs in **Blown Away**.

Curious	★★★
Connect	★
Collaborate	★★★
Communicate	★★★
Create	★★★
Consider	★

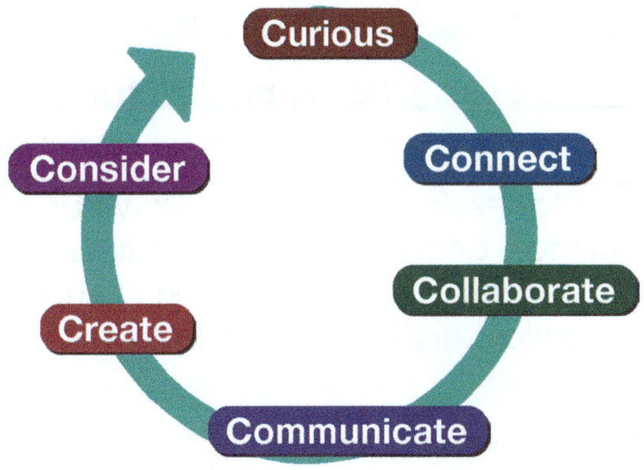

Problem-solving heuristics

The grid shows which problem-solving strategies children are likely to use when investigating the **Connect** tasks in **Blown Away**. These strategies are further exemplified in the **Connect** stage below (page 157).

Draw a diagram	✓
Make a table / spot a pattern	✓
Make a model / act it out	
Try then improve	
Create a systematic list	
Solve a simpler but related problem	
Work backwards	

Key mathematical ideas

This task provides an opportunity for children to investigate forces alongside mathematics. Children will explore how to make a paper hovercraft and then consider how they can improve their design. The first challenge is to decide how they are going to evaluate their designs. For example, is it more important for the hovercraft to be the fastest, travel in the straightest line or cover the greatest distance? The criterion selected must be quantifiable (measurable).

Once they have chosen an evaluation criterion, they can start to consider the variable that they are going to change. These are likely to be:

- construction material (for example, aluminium foil, card, newspaper, sugar paper);
- size of original sheet of paper.

It is important that the children understand the importance of only changing one variable. Children cannot change both size and construction material as that would make it difficult to identify which variable was responsible for the change in outcome.

Speed travelled is easy to calculate using the formula: **Speed** = **Distance** ÷ **Time**. You may wish to provide calculators for this.

Children will need to identify how they are going to collect their data. Which measurements are important and what will their recording table look like?

Once they have completed their data collection, children will need to translate the tabulated information into a graphical form (bar chart, pictogram, and so on). By interpreting their graphs, they will be able to identify how their variable affects their hovercraft and which was the 'best'. Ultimately the children will be presenting why their design should be selected, using all of the data they have collected.

Answers are not provided, as the focus of this task is on how children carry out their investigations and present their results.

Teaching guidance

If you are using the **Express Pathway**, please turn to pages 156–159.

Teaching guidance for the **Full 6C Pathway** is given below.

The **Teacher's Presentation** PowerPoint file has slides specifically mentioned in this teaching guidance. Some of the slides include notes (within the PowerPoint file) providing 'stumble support' suggestions, that may help you to support children in their problem-solving thinking and enable access for children who are less confident.

Awe and wonder

Additional resources required: square sheets of paper

Slide 1

Display slide 1 to introduce the task. Explore what the children already know about hovercrafts. If needed, share a video of a hovercraft.

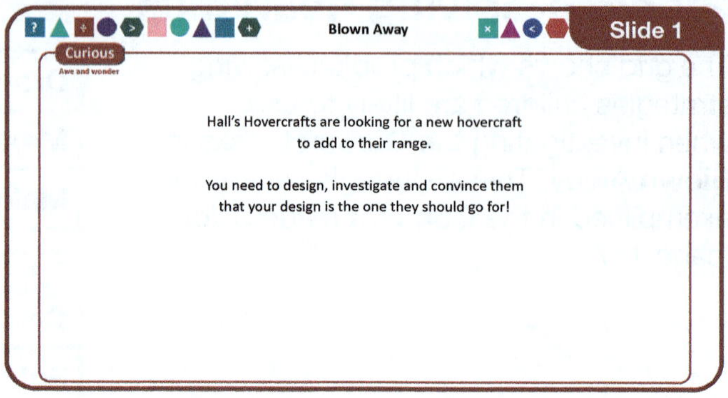

Slides 2 and 3

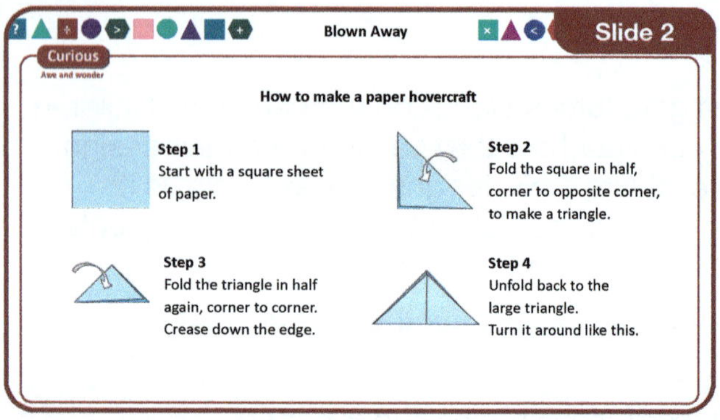

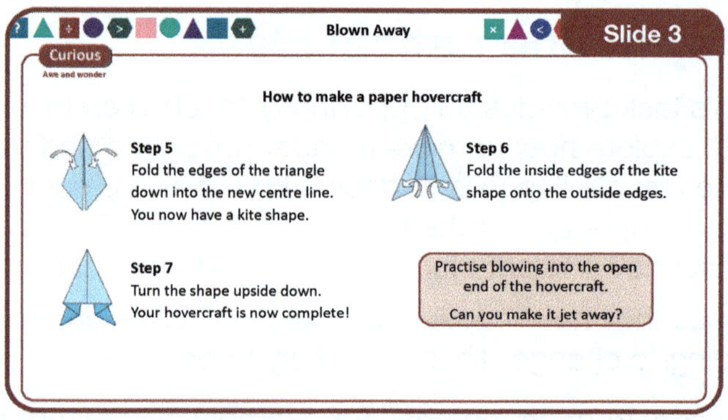

Display slides 2 and 3 one at a time. (Alternatively, you could print these two slides and distribute to children.)

Show children how to follow the seven steps to fold the paper hovercraft.

Give them time to make their own hovercrafts and play with them for a short period of time.

- I have noticed that it doesn't travel in a straight line.
- I can see that the harder I blow, the further it travels.
- I have noticed that when … blows, his / hers goes further.
- I have noticed that where you blow from makes a difference.
- I can see that it travels further on the smooth floor than on the carpet.

⚠ It is important that children notice that how hard you blow affects the travel of the hovercraft. They will need to take account of this when doing their investigation.

Develop and deepen

Slide 4

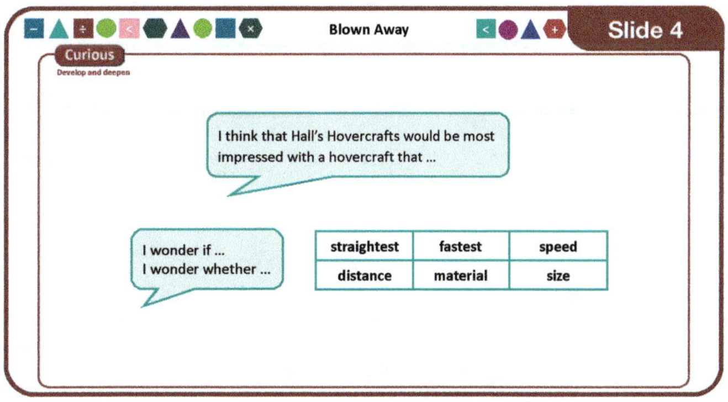

Display slide 4.

Present children with the sentence starter:

- I think that Hall's Hovercrafts would be most impressed with a hovercraft that …

Discuss a few ideas.

Also, ask the children to think about what they want to compare and change between different hovercraft models, and how they can measure that.

Encourage children to use the words in the word bank to develop and deepen their thoughts and questions.

straightest	**fastest**	**speed**
distance	**material**	**size**

Possible lines of enquiry:

- I wonder whether the type of paper affects the hovercraft.

Possible lines of enquiry linked to the word bank:

- I wonder if the size of the sheet of paper affects the hovercraft.
- I wonder if the size of paper affects how far the hovercraft travels.
- I wonder if the material used affects how fast the hovercraft is.

Possible lines of enquiry that children may raise now, but should be held back until the **Create** task:

- I wonder if there is an ideal material and size.
- I wonder if I can make a hovercraft using a different design.

Consider whether / how to collate and display all the suggested questions, and whether / how to steer children towards selecting an appropriately challenging question.

Pupil Resources for Full and Express Pathways

Distribute the **Pupil Resources** for **Blown Away** to the children to help support them in their exploration of the different tasks. Use either of the **Curious** **Pupil Resources** depending on whether you are following the **Full Pathway** or the **Express Pathway**. The **Pupil Resources** for the remaining 5Cs are common to both pathways.

To provide additional flexibility in the classroom, each of the **Pupil Resources** is also provided in the **Teacher's Presentation** PowerPoint file (see slides 5–8), so you can display these at any time if you wish.

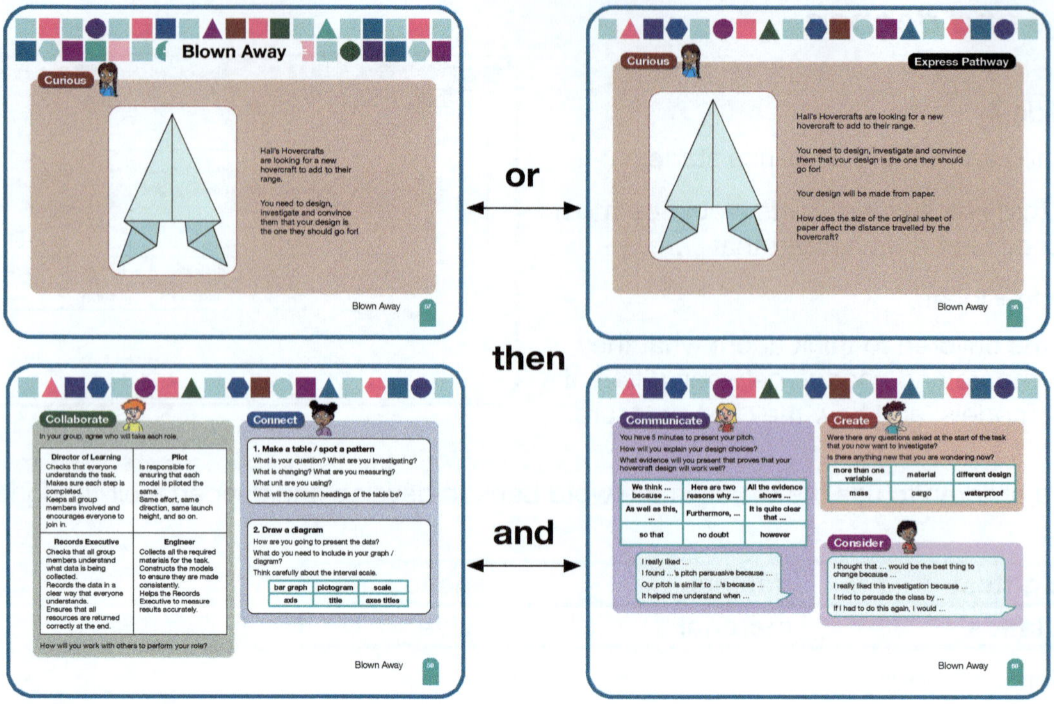

Express Pathway

If you are following the **Express Pathway**, spend a few minutes discussing the **Curious Pupil Resource** with the children. (This is slide 6, so you could display it if you wish.) Use some of the prompts on the previous page to help.

Collaborate

Refer to the **Collaborate** task of the **Pupil Resource**.

Slide 9

Start by splitting the class into groups of four. If the number will not divide evenly, children will have to take on multiple roles, for example, the engineer and pilot can be taken on by the same child.

Print slide 9 and distribute a copy to each group. This is an ideal opportunity to explore the impact of assigning jobs to different members of the group:

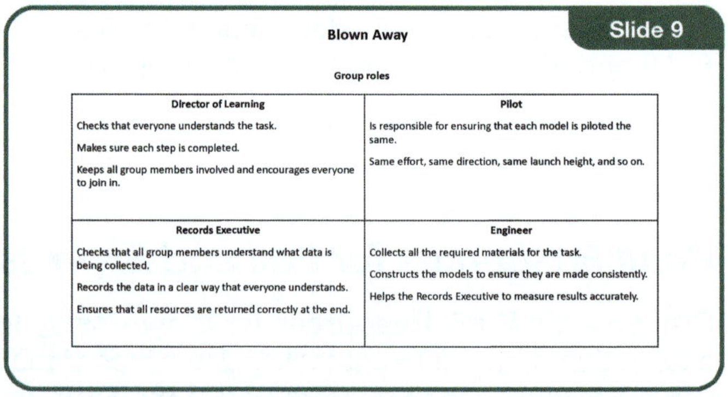

- Director of Learning;
- Pilot;
- Records Executive;
- Engineer.

Once roles have been assigned, groups need to consider the name of their company.

For example, they could consider:

- a combination of their names, for example, FozEmmHarKab (Fozia, Emma, Harper, Kabir);
- their initials, for example, FEHK Industries;
- rhyming names, for example, Refine Design or Improve and Prove.

Two tasks involving different strategies are suggested for children to make connections.

Knowing your class, you might decide to assign each of the below tasks to a particular role or ask the children to work on all the tasks as a group.

1. Make a table / spot a pattern

Slide 10

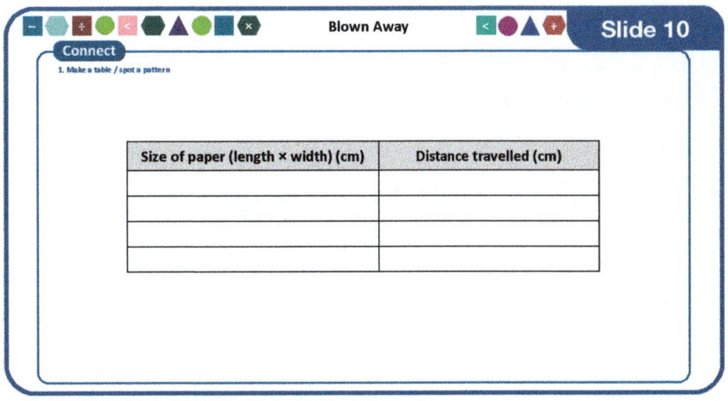

Children will need an organised way to collate their information and document their findings. Encourage them to think about what they are measuring to help them compare hovercrafts and which unit of measurement they will be using.

Display slide 10, if needed, to show example column headings that could be used when investigating the question, 'How does the size of paper affect how far the hovercraft travels?'

Groups will need to consider which graphical representation will convey their data most effectively.

2. Draw a diagram

Slide 11

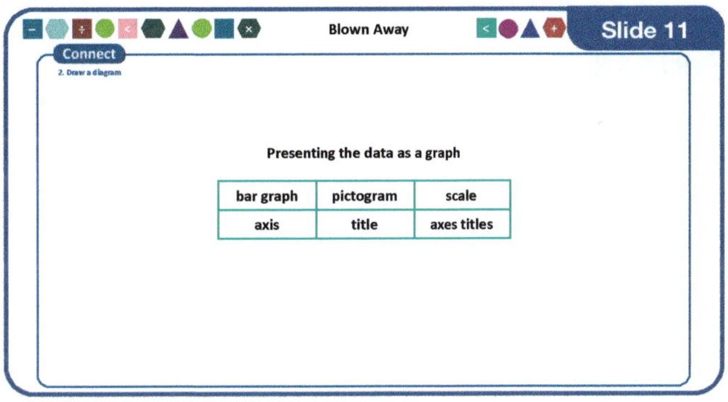

Display slide 11.

Encourage children to think about which graph or diagram will best represent their data and their findings.

- *Which types of graphs or diagrams have you come across before?*
- *What are the benefits of each type?*
- *What are the disadvantages of each type?*
- *Have you thought about the intervals - what will happen if you use smaller or larger intervals?*

bar graph	**pictogram**	**scale**
axis	**title**	**axes titles**

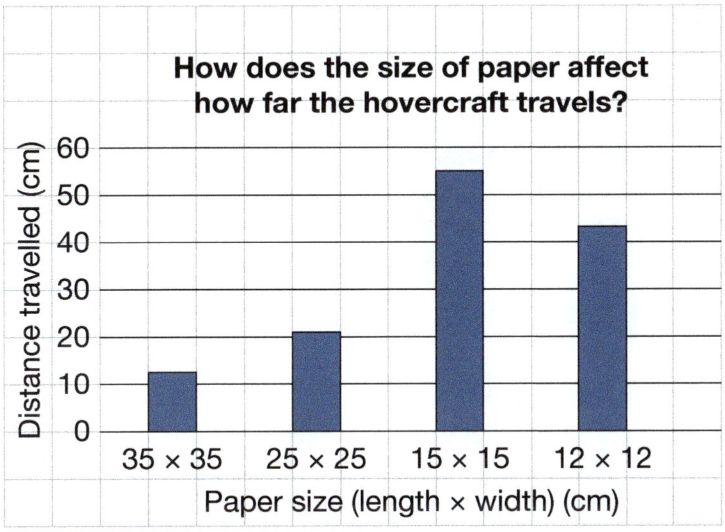

An example bar graph

⚠ Keep an eye on the scales that the children are using. Are they appropriate? Are the intervals equal? Also, check to see if they have included all the relevant information (for example, all of their data points, main title, axes titles).

Communicate

Refer to the **Communicate** task of the **Pupil Resource**.

Slide 12

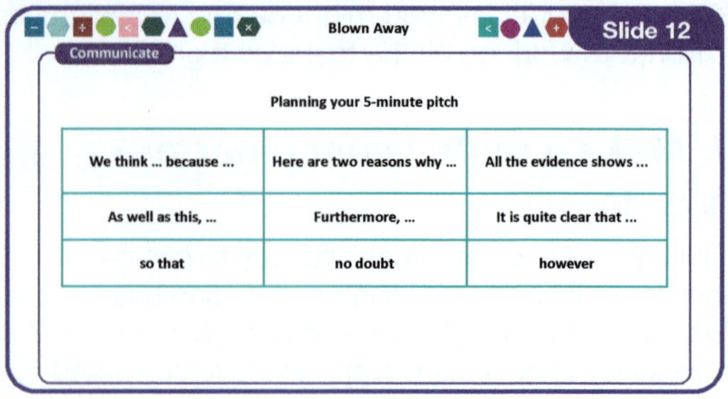

Display slide 12.

This is an opportunity for children to engage in an entrepreneurial experience and develop a pitch based on their hovercraft design.

Their pitch must be a maximum of 5 minutes. Children should consider the following:

- *How are you going to introduce yourselves?*
- *How will you explain your design choices?*
- *What evidence will you present that proves that your hovercraft design will work well?*
- *Why is your idea to improve the hovercraft better than other groups?*
- *Are there any downsides to your improvements?*

We think … because …	**Here are two reasons why …**	**All the evidence shows …**
As well as this, …	**Furthermore, …**	**It is quite clear that …**
so that	**no doubt**	**however**

Slide 13

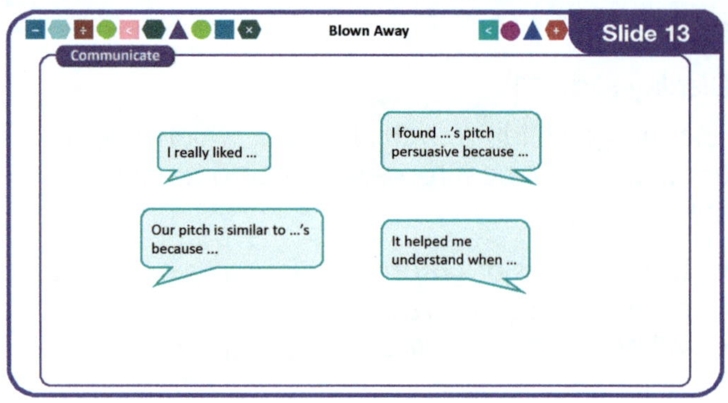

Display slide 13.

Encourage children to compare each other's pitches, using the sentence stems on the slide if needed.

Create

Refer to the **Create** task of the **Pupil Resource**.

Slide 14

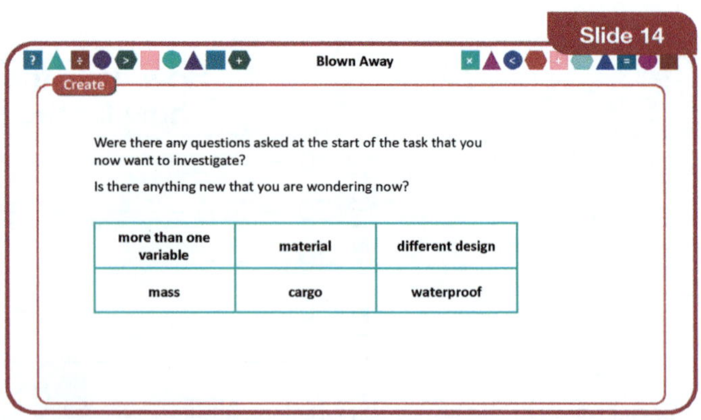

Display slide 14.

- *Were there any questions asked at the start of the task that you now want to investigate?*
- *Is there anything new that you are wondering now?*

Children can use the word bank to prompt further questions of their own.

more than one variable	**material**	**different design**
mass	**cargo**	**waterproof**

- I wonder if the hovercraft would travel further if we changed the size of the paper and the material.
- I wonder if there is an ideal material and size.
- I wonder if I can make a hovercraft using a different design.
- I wonder which design of hovercraft would be 'best' at carrying cargo.

⚠ Ensure children establish and clarify how they are going to define 'best'.

Refer to the **Consider** task of the **Pupil Resource**.

Slide 15

Display slide 15.

Encourage children to think about the strategies that they have used. The prompts below may help get children started.

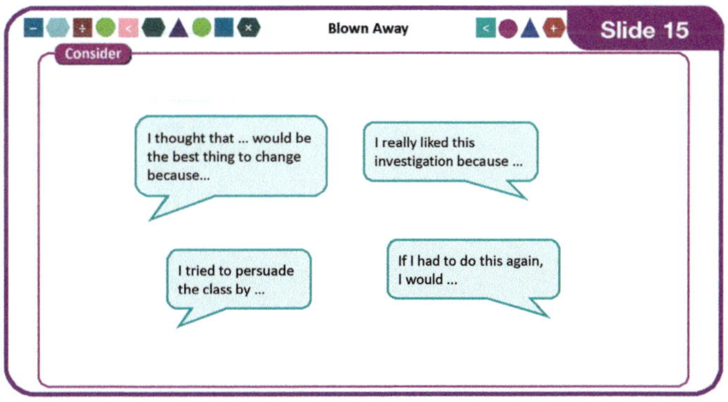

Cognitive learning domain

- I thought that … would be the best thing to change because …
- Next time, I will …
- I noticed that ...

Affective domain

- I really liked this investigation because …
- My 'aha' moment was …
- I tried to persuade the class by ...

Metacognitive domain

- I found … helpful.
- I found … unhelpful.
- If I had to do this again, I would …
- I organised my work ...